Heterogeneous Constructions

Heterogeneous Constructions

Studies in Mixed Material Architecture

Aaron Forrest
Brett Schneider
Yasmin Vobis

Birkhäuser
Basel

p 38

p 44

p 62

p 68

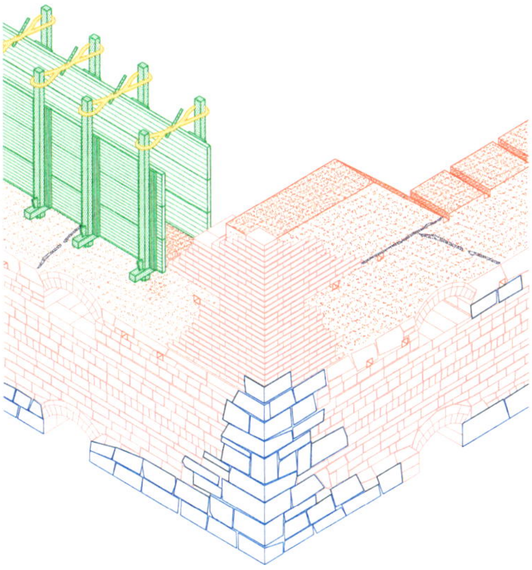

p 50

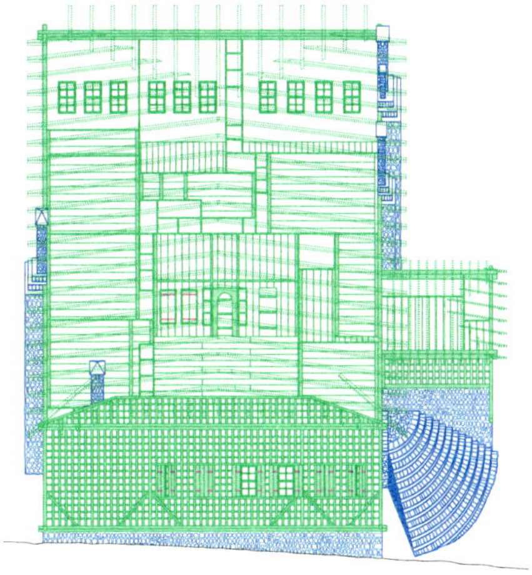

p 56

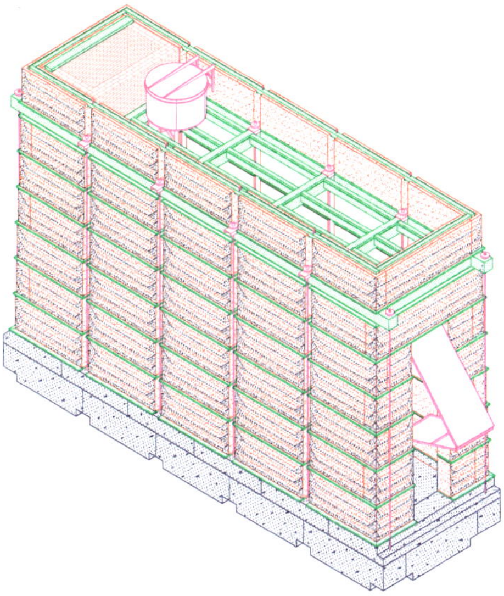

p 74

p 80

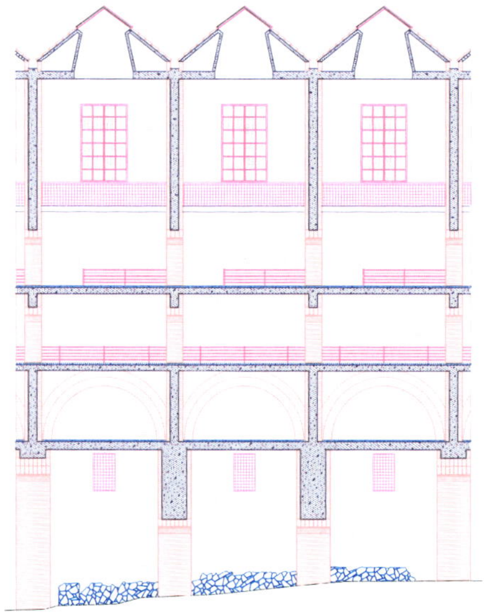

p 86

p 92

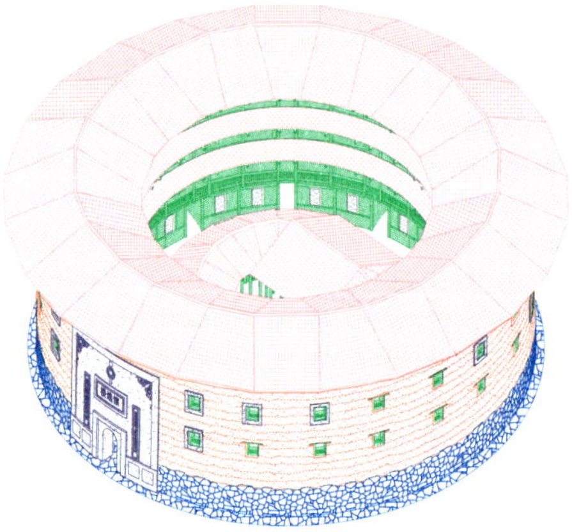

p 110

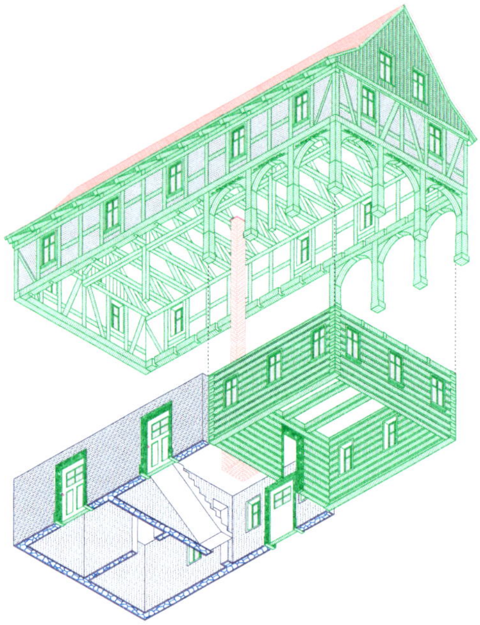

p 116

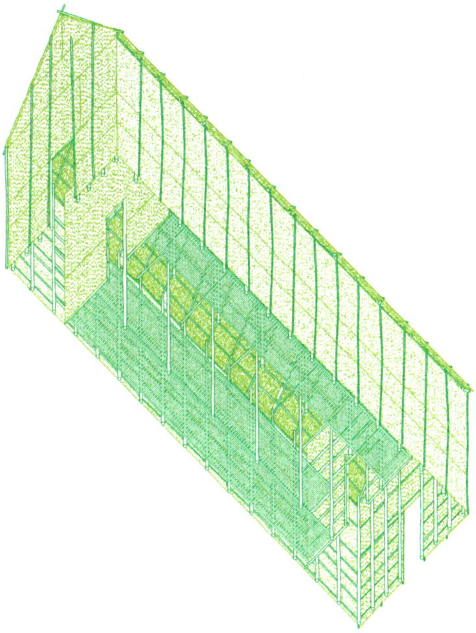

p 98

p 104

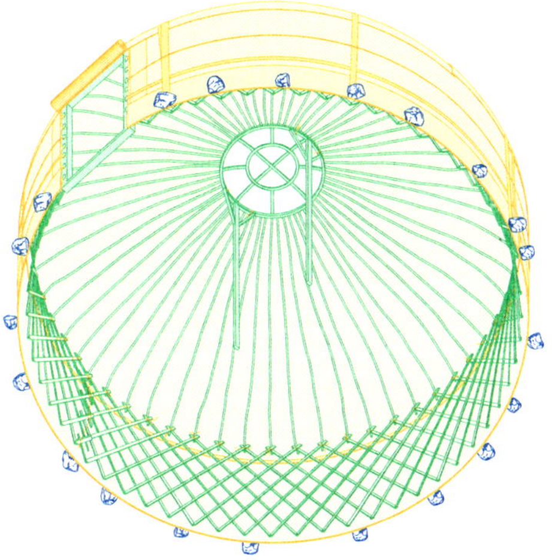

p 122

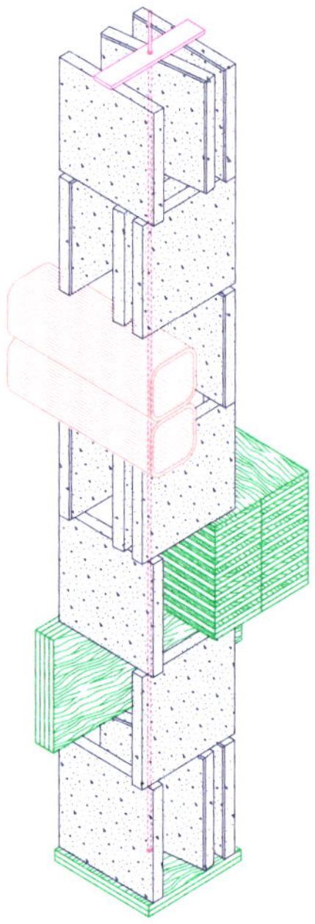

p 172

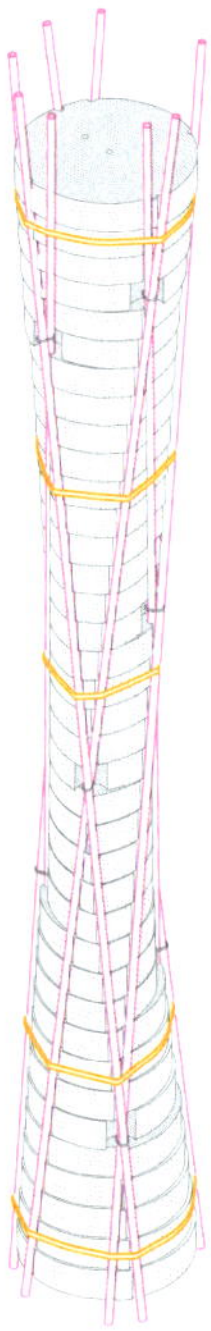

p 176

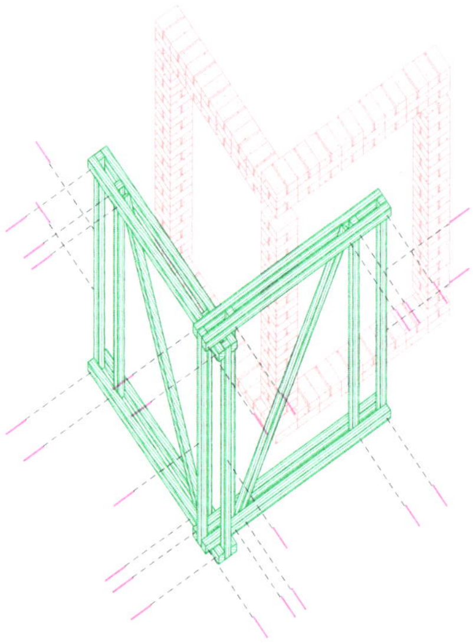

p 188

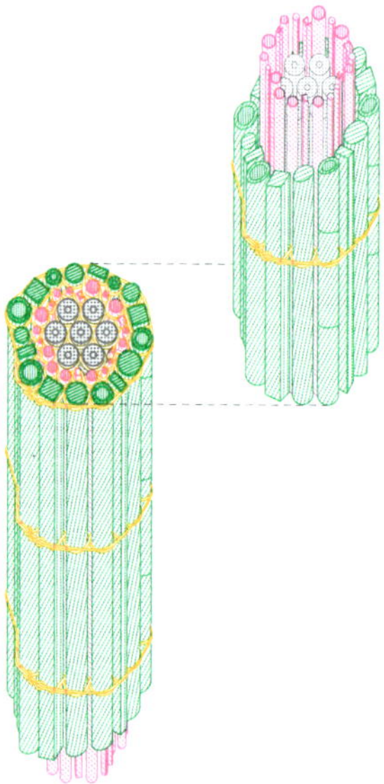

p 192

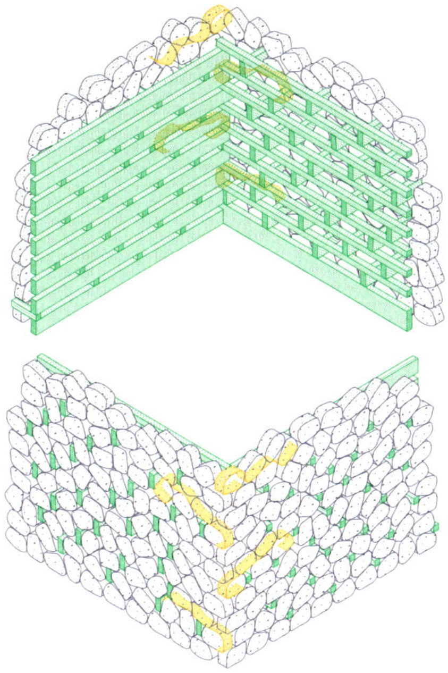

p 180

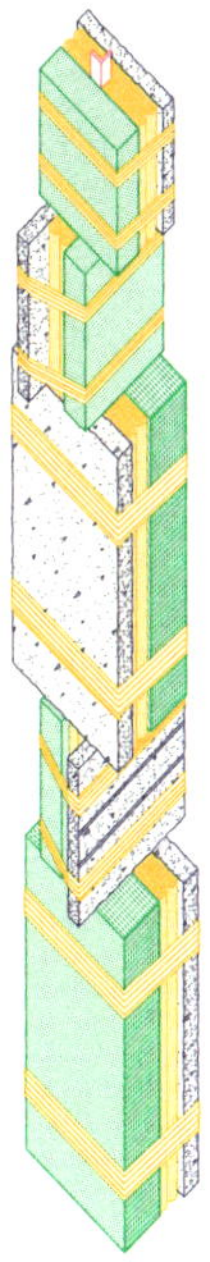

p 184

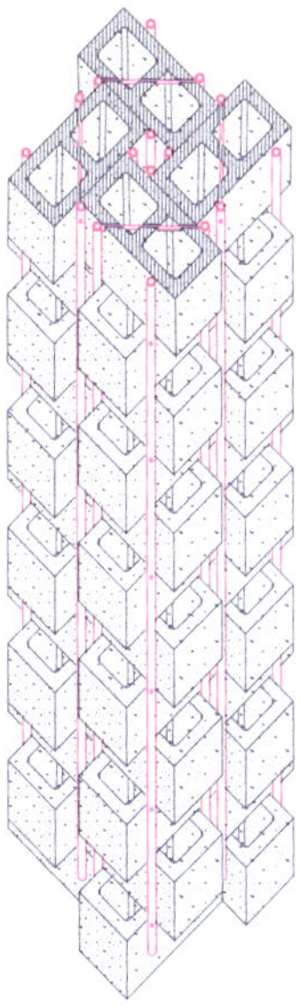

p 196

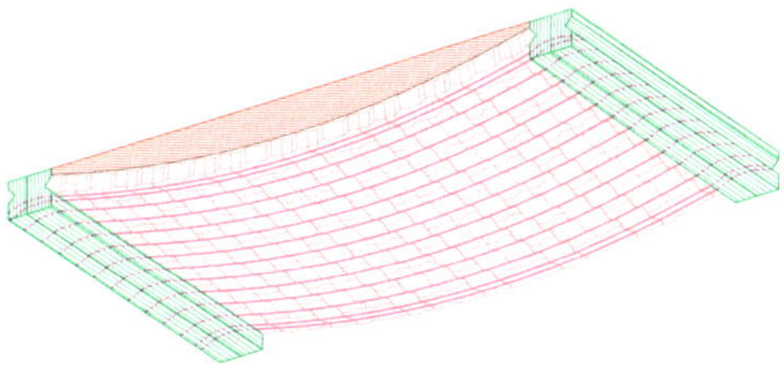

p 200

Ceramic	and	Concrete
Ceramic	and	Concrete
Ceramic	and	Earth
Ceramic	and	Earth
Ceramic	and	Metal
Ceramic	and	Metal
Ceramic	and	Metal
Ceramic	and	Stone
Ceramic	and	Wood
Ceramic	and	Wood
Ceramic	and	Wood
Ceramic	and	Wood
Concrete	and	Ceramic
Concrete	and	Ceramic
Concrete	and	Metal
Concrete	and	Metal
Concrete	and	Metal
Concrete	and	Metal
Concrete	and	Textile
Concrete	and	Wood
Concrete	and	Wood
Concrete	and	Wood
Concrete	and	Wood
Earth	and	Ceramic
Earth	and	Ceramic
Earth	and	Metal
Earth	and	Metal
Earth	and	Stone

as	Collage	p 172
as	Sandwich	p 86
as	Scaffold	p 50
as	Stack	p 200
as	Collage	p 172
as	Fill	p 80
as	Stack	p 200
as	Scaffold	p 50
as	Collage	p 172
as	Fill	p 104
as	Fill	p 188
as	Scaffold	p 50
as	Collage	p 172
as	Sandwich	p 86
as	Collage	p 172
as	Collage	p 92
as	Sandwich	p 196
as	Scaffold	p 184
as	Scaffold	p 184
as	Collage	p 172
as	Collage	p 92
as	Scaffold	p 180
as	Scaffold	p 184
as	Scaffold	p 50
as	Stack	p 200
as	Stack	p 200
as	Stack	p 74
as	Collage	p 116

Earth	and	Stone
Earth	and	Wood
Earth	and	Wood
Earth	and	Wood
Earth	and	Wood
Earth	and	Wood
Earth	and	Wood
Metal	and	Ceramic
Metal	and	Ceramic
Metal	and	Ceramic
Metal	and	Concrete
Metal	and	Concrete
Metal	and	Concrete
Metal	and	Concrete
Metal	and	Earth
Metal	and	Earth
Metal	and	Other
Metal	and	Other
Metal	and	Textile
Metal	and	Textile
Metal	and	Textile
Metal	and	Wood
Metal	and	Wood
Metal	and	Wood
Metal	and	Wood
Metal	and	Wood
Other	and	Metal
Other	and	Metal

as	Scaffold	p 50
as	Collage	p 116
as	Sandwich	p 196
as	Scaffold	p 50
as	Scaffold	p 62
as	Stack	p 74
as	Wrap	p 110
as	Collage	p 172
as	Fill	p 80
as	Stack	p 200
as	Collage	p 172
as	Collage	p 92
as	Sandwich	p 196
as	Scaffold	p 184
as	Stack	p 200
as	Stack	p 74
as	Wrap	p 176
as	Wrap	p 192
as	Scaffold	p 184
as	Wrap	p 176
as	Wrap	p 192
as	Collage	p 172
as	Collage	p 92
as	Scaffold	p 184
as	Stack	p 74
as	Wrap	p 192
as	Wrap	p 176
as	Wrap	p 192

Other	and	Textile
Other	and	Textile
Other	and	Wood
Plants	and	Textile
Plants	and	Wood
Plants	and	Wood
Stone	and	Ceramic
Stone	and	Earth
Stone	and	Earth
Stone	and	Wood
Stone	and	Wood
Stone	and	Wood
Stone	and	Wood
Textile	and	Concrete
Textile	and	Metal
Textile	and	Metal
Textile	and	Metal
Textile	and	Other
Textile	and	Other
Textile	and	Plants
Textile	and	Wood
Textile	and	Wood
Textile	and	Wood
Textile	and	Wood
Textile	and	Wood
Wood	and	Ceramic
Wood	and	Ceramic
Wood	and	Ceramic

as	Wrap	p 176
as	Wrap	p 192
as	Wrap	p 192
as	Sandwich	p 98
as	Sandwich	p 98
as	Wrap	p 38
as	Scaffold	p 50
as	Collage	p 116
as	Scaffold	p 50
as	Collage	p 116
as	Fill	p 56
as	Scaffold	p 50
as	Stack	p 68
as	Scaffold	p 184
as	Scaffold	p 184
as	Wrap	p 176
as	Wrap	p 192
as	Wrap	p 176
as	Wrap	p 192
as	Sandwich	p 98
as	Sandwich	p 98
as	Scaffold	p 184
as	Scaffold	p 44
as	Wrap	p 122
as	Wrap	p 192
as	Collage	p 172
as	Fill	p 104
as	Fill	p 188

Wood	and	Ceramic
Wood	and	Concrete
Wood	and	Concrete
Wood	and	Concrete
Wood	and	Concrete
Wood	and	Earth
Wood	and	Earth
Wood	and	Earth
Wood	and	Earth
Wood	and	Earth
Wood	and	Earth
Wood	and	Metal
Wood	and	Metal
Wood	and	Metal
Wood	and	Metal
Wood	and	Metal
Wood	and	Other
Wood	and	Plants
Wood	and	Plants
Wood	and	Stone
Wood	and	Stone
Wood	and	Stone
Wood	and	Stone
Wood	and	Textile
Wood	and	Textile
Wood	and	Textile
Wood	and	Textile
Wood	and	Textile

as	Scaffold	p 50
as	Collage	p 172
as	Collage	p 92
as	Scaffold	p 180
as	Scaffold	p 184
as	Collage	p 116
as	Sandwich	p 196
as	Scaffold	p 50
as	Scaffold	p 62
as	Stack	p 74
as	Wrap	p 110
as	Collage	p 172
as	Collage	p 92
as	Scaffold	p 184
as	Stack	p 74
as	Wrap	p 192
as	Wrap	p 192
as	Sandwich	p 98
as	Wrap	p 38
as	Collage	p 116
as	Fill	p 56
as	Scaffold	p 50
as	Stack	p 68
as	Sandwich	p 98
as	Scaffold	p 184
as	Scaffold	p 44
as	Wrap	p 122
as	Wrap	p 192

Visual Index 4
Verbal Index 10

Examples 37

Prototypes 171

Contents

Foreword 20
Making Do
Jesús Vassallo
Introduction 24
Aaron Forrest
Brett Schneider
Yasmin Vobis
Glossary 32

Provocations 129
Category Errors
Aaron Forrest
The Difficult Imperfect Whole
Jeannette Kuo
Mixed Media
Yasmin Vobis
Grid, Machine, Mud, Work, Air
Ajay Manthripragada
How to Build Heterogeneously
Brett Schneider

See Also 207
Bibliography 217
Image Credits 221
Biographies 222
Acknowledgments 223
Credits 224

Making Do

Jesús Vassallo

Looking out the window today in almost any large global city, it can actually be quite difficult to tell exactly where we are. The homogeneity of building types, construction materials, and the blandness of planning principles all contribute to a certain morass of structure and infrastructure, a condition that we have come to accept as pervasive, perhaps inevitably defining of contemporary urban life.

It is then easy to forget, in our current predicament, that until relatively recently the business of building cities and buildings was a deeply regional affair. Historically, construction materials were chosen based on local availability of resources, the difficulty of their extraction and processing, and the rate of their natural replenishment. Places near forests built with wood, while places with clay soils and no forests built with bricks. Even the building trades were regionally distinct, with names for the same building materials or tools varying from valley to valley, just a few miles apart.

It was not until the advent of industrialization that construction started a process of gradual homogenization. Even then, the link between the extraction of resources and their deployment remained visible for some time. For instance, the early iron ore mining and smelting industry in eighteenth-century Germany gave birth to a land management system aimed at the sustainability of its forests. By establishing yearly amounts of timber to be felled and the proportions that could be reserved for industry, construction, and other applications, it gave way to a variation of the now revered half-timber construction system. Here the need to save wood as much-prized fuel for the iron ore business resulted in a construction logic in which rubble, brick, and other earth-based materials were added to the wall in order to reduce the amount of timber needed in a structure.

Along the way of its development, the process of industrialization produced a series of technological innovations—the railroad, the combustion engine, the shipping container—that collectively enabled the emergence of a regime of frictionless transportation of goods and raw materials at the scale of the globe. Only then did it become possible to conceptually dislocate the points of origin and application of construction materials. Not only could all buildings be constructed in concrete and steel in an international style, but the dwellers of those buildings could remain oblivious to the location of the mines, quarries, kilns, and furnaces necessary for their construction.

The twin processes of industrialization and globalization sketched here underwrite our moment of late global capitalism, with its parallel logics of extraction and abstraction. In its pursuit of a dream of homogenous affluence over two centuries, this system conveniently ignores what we could describe as a series of hidden costs, namely that its greenhouse emissions are tipping our planet's climate out of kilter while depleting life-sustaining resources as if there were no tomorrow.

As architects, we wake up today to the sobering realization that our ways are untenable, that we must simultaneously reverse the contributions of the construction industry toward global warming and adapt our buildings and cities to withstand the liabilities of our already changed climate. The decarbonization of our industry requires nothing short of a complete change of paradigm, comparable in scale and scope to the wholesale transformation of construction practices undertaken between preindustrial times and our own time.

Granted, it is not possible or desirable to turn back the clock on progress, but our crisis requires an all-hands-on-deck approach in which no individual contribution should be ignored, including the recovery and revaluation of materials or techniques that were once discarded as primitive. Part of our work in adopting a postcarbon paradigm is then to figure out how we can reintroduce a more localized logic to our construction practices and make do with the available resources at hand.

Inevitably, this new low-carbon local paradigm will lead to a proliferation of unexpected mixes and assemblies between high tech and low tech, between organic and inorganic materials. This will require changes in our standards and the way in which we evaluate construction. Already, as we work to replace petroleum-based insulators with plant-based solutions, we can see the emergence of a renewed tolerance for imperfection in the appreciation of materials. Who would have told us a few years ago that Western architects would seriously consider straw or hemp as viable insulation materials?

Similarly, the renewed interest in techniques such as rammed earth or stone construction presents a timely opportunity for developing nations to teach industrialized ones. Since some countries were never able to fully buy into the industrial construction complex, they did not completely lose their knowledge and craftsmanship of traditional materials and can now lead the way in their reintroduction.

We must however avoid falling into a medieval fantasy of pure, handmade, passive buildings. The scale and urgency of change means that the future of construction will be hybrid rather than exclusive, as we strive to adopt a wide array of techniques at scale in an effort to move the needle. This inclusive approach will likely spell an aesthetic of brinkmanship different from the sleekness of modernism, a logic of smart heterogeneity instead of total integration.

Already, the need to operate on existing building stock as opposed to systematically demolishing and rebuilding from scratch is resulting in a retooling of our everyday practices. In this new normal, every project is at its starting point a hybrid, a mediation between the construction logics of different time periods. This is relevant because, despite being in desperate need of radical reform, our planet is already fully urbanized, which calls for an innovative and heterogenous approach able to both valorize and transform existing conditions.

As we step into our messy future, it is useful to remember that construction has always been a heterogenous affair, each building an assemblage of a multitude of elements and materials requiring as many decisions about their selection and combination. It is important then that we reexamine the system of values driving those decisions, most of all the notion of efficiency that has traditionally reigned supreme in the minds of architects, engineers, and builders. Hopefully, we will accomplish this goal by embracing a much-expanded notion of what things cost, not just to us individually in the moment but to everyone in the long term.

Introduction

Aaron Forrest
Brett Schneider
Yasmin Vobis

In Sigfried Giedion's 1928 manifesto of industrial construction, *Building in France, Building in Iron, Building in Ferroconcrete*, an unusual pair of captioned images appears unreferenced in the main text: a framing elevation and plan of the Menier Chocolate Factory in France, designed and built in 1871–72.[Fig 1] The image of the factory is unlike those that surround it: it is one of the few buildings in the first half of the book (which almost exclusively highlights engineering marvels of the previous century); and was designed collaboratively by architect Jules Saulnier and structural engineer Armand Moisant, rather than either alone. Giedion's caption announces the building as the "first thoroughly consistent IRON-SKELETON BUILDING," notable both for its self-supporting iron frame as well as the uniquely flat facade that the structural system enables.[1]

The information that Giedion omits is just as notable. A closer look at the building itself reveals a highly decorative polychrome brick facade coplanar with the iron frame, invisible in the framing elevation and unmentioned in the captions.[p 80] The very masonry that enables the structure's apparent abstraction in reality presents a structural and aesthetic ambiguity that belies Giedion's entire premise. The facade's planarity, achieved by filling the voids in the lightweight iron frame with brick—itself tinted and patterned to create the illusion of decorative depth—likewise suggests the impossibility of separating the structural roles of the two materials. The brick lends mass—and likely stiffness—to a frame that is both unusually light and imperfectly triangulated. Similar to the seismic properties of early skyscrapers,[2] the relationship between masonry and iron in the factory is difficult to quantify but real nonetheless.

The Menier Chocolate Factory points to a path not taken by modern architecture. Giedion's reading of the building as a whole is in service of an ideology predisposed to both the separation of functions and abstract aesthetics. But the building's richness lies not in the transition to the isolated frame, but in the interaction of iron and brick toward the multiple functions of structure, enclosure, and decoration within a single, inseparable plane of construction.

Giedion sees only the frame and misses the totality of the building. The arc he favors toward functional discreteness and geometric abstraction is only one possible trajectory for the burgeoning modern movement. That arc is today well established as the dominant force within modernism, but it was accompanied in lockstep by a similar tendency within the construction industry and building science toward the functional separation of layers in every building assembly, today exemplified by the much-cited "Perfect Wall" proposed by Joseph Lstiburek. Here the exterior wall is represented as a series of layers from cladding to control layers to structure,[Fig 2] and, as the author notes, the wall "works everywhere in all climate zones."[3] It is this discrete layering and placelessness that is the unconscious given in much contemporary construction. From Villa Savoye to nearly every building constructed in the West today, the separation of functionality—and the abstraction of this function—has been consistently reinforced over the course of a hundred years of architectural and construction history.

This book is an attempt to chart an alternative course: one based in heterogeneous construction. This very concept is, at first, somewhat elusive. On a fundamental level, heterogeneity could describe nearly all of modern construction: most buildings today are combinations of different materials selected for their distinct performances. Despite their assembly into an apparently unified whole, the role of each of these layers and materials is typically independent of the others—that is to say, they are heterogeneous *as a collection*, but not in the way they perform *collectively*.

We look here instead to constructions where the materials employed take on multiple roles in relation both to one another and to the assembly as a whole, interacting in complex and sometimes ambiguous ways. In heterogeneous construction, multiple materials may operate as structure, envelope, and decoration in ways that are nondiscrete, but that in conjunction do more than any one material may be able to on its own. In this manner, an emphasis is placed on relationships between materials as much as the characteristics of individual materials themselves.

Heterogeneous construction accepts from the start that buildings are assemblies of many materials, parts, and systems. We seek to understand how these parts have worked together through an examination of architectural examples from a wide range of global building cultures and epochs, looking far beyond the typical entries in the modernist canon. These include a wide swath of vernacular architecture, including many examples from the Global South, as well as proto-industrial buildings of the nineteenth and early twentieth centuries, of which the Menier factory is but one. The large majority emerge from empirical building traditions driven by craft and material availability in the case of vernacular construction, or the experiments around the introduction of iron and reinforced concrete during the Industrial Revolution. The case studies likewise include contemporary examples that, following the conventions of modern construction, are the result of discrete but relational material strategies. In these buildings, a rational combination of materials allows for an abstract understanding of function as a whole. But the sum of the parts results in a more complex performance and often relies on empirical methods in addition to analytical methods in order to be developed.

Fifteen examples form the heart of the book, alongside essays from both the authors and outside contributors, and design experiments that explore the contemporary potential for heterogeneous construction. The case studies are examined in-depth through photographs and full-color analytical line drawings in which materials are color-coded as a visual accounting of their heterogeneity. Our research began by casting a very wide net, and proceeded by narrowing progressively to a smaller set of representative examples from disparate eras and building cultures.[4] The aim was to illustrate a set of material relationship types that could be seen to bridge divisions of time and place: Stacks, Sandwiches, Fills, Wraps, Collages, and Scaffolds.[5] The examples are purposely mixed, juxtaposing types and origins as a way of allowing the immersed reader a comparative understanding while simultaneously demonstrating the breadth of possibilities made available by heterogeneous construction.

Our purpose here, though partly historical in scope, is not atavistic. On the contrary, the point of looking back is, as it has been so eloquently posed by Vittoria di Palma,[6] to be confronted by the radical potential of cultural artifacts that fall far outside the realm of accepted contemporary practice. The revaluation of heterogeneity that these examples engender suggests new avenues for contemporary construction. Notably, it creates the possibility for more sustainable construction methods by allowing a more consequential inclusion of nonindustrial, locally sourced, or reused materials in architecture. By considering materials in combination rather than in parallel monofunctional layers, there are likewise opportunities for increased efficiencies and performance across the assembly. One simply need consider the invention of reinforced concrete to understand the effectiveness of such a complementary material approach. Finally, and perhaps most consequential from the point of view of architecture as cultural production, the preponderance of historical examples demands that we look beyond a still dominant mode of modernism for alternative approaches to construction and design to address the important challenges of our time.

Any serious foray into heterogeneous construction as a contemporary concern must confront a set of deeply challenging questions: cultural, creative, ethical, and practical. Many of the examples studied were developed empirically over generations and their exact performance would be difficult to replicate, rationalize, or calculate using modern analytical methods. Regulatory approvals, which are based on these analytical methods, would be difficult to attain unless extensive testing of the proposed construction type is available in addition to calculations. Many of the constructions further depend on labor conditions and material ecologies limited to specific places and, in some cases, historical eras. As a result, any contemporary approach to material heterogeneity must be reimagined through contemporary construction standards, engineering processes, material availability, and cultural and labor conditions.

These issues demand serious exploration in their own right. The framing essays in this volume are overlapping responses to a series of open questions. How can we make sense of the diversity of heterogeneous constructions while finding common threads to build on? How can we evaluate, quantify, and qualify the complexity of indiscrete heterogenous construction in order to justify building in our current construction and regulatory environment? What are the theoretical implications of these heterogeneous constructions and how does this inflect our more general design thinking? Contributions from scholar-architects Ajay Manthripragada and Jeannette Kuo provide deep forays into broader disciplinary, cultural, social, and ecological questions of the heterogeneous through the analysis of specific buildings, practices, and theories both historical and contemporary. These essays are explorations and, for the most part, leave the questions they ask unresolved. For us, this irresolution is a precondition to research as designers. Rather than a search for universal truth, this type of inquiry seeks contingent and multiple solutions to the multivariable problem of architecture and construction.

The work presented in the Prototypes section explores similar questions to those posed in the essays, but via the nimble medium of speculative design. These prototypes, designed and built by students from three universities over two different time periods, seek to extrapolate on concepts from the case studies through the invention and construction of full-scale, heterogeneous building elements. The prototypes do not revive historical construction techniques, but instead seek new definitions of heterogeneous construction appropriate to our own context: contemporary, off-the-shelf, upcycled, and discarded construction materials. Situated somewhere in the gray area between construction mock-up and 1:1 scale model, the prototypes seek to be both performative and decorative, and to suggest a trajectory for how designers might rethink the combination of basic industrial building components toward alternative ends.

Can heterogeneous construction exist in a contemporary context defined by the wide availability of cheap, mass-produced construction materials, a climate-driven imperative for reuse, and scarcity of craft labor? Looking beyond the prototypes, certain contemporary practices discussed in these pages—Roger Boltshauser, Atelier De Vylder Vinck Taillieu, and the TUM Building Simply group, among others—are already hinting at how a technically rigorous and culturally situated approach to the heterogeneous might be implemented in a modern construction environment. The approaches of these designers are inherently cross-disciplinary and typically make use of design and construction processes that deploy a strategic combination of analytical tools and empirical testing. This combined approach recalls the practices of the engineer-builders of the nineteenth and twentieth centuries without suggesting a revival of the master-builder paradigm. Like heterogeneous construction more generally, they imply a far closer dialogue between architect, engineer, and builder than is commonplace today as a mechanism to reconnect questions of culture and design to the craft of building.

Between the historical case studies and contemporary experiments presented here, a dialectical relationship emerges. The drawings present a view to a rich and diverse set of building traditions that is at once immersive and technical, with unanticipated lessons to be drawn and the potential to provide the basis for contemporary construction techniques yet to be invented. As a discipline, architecture has an imperative to creatively address the twenty-first century's crises of carbon, sustainability, reuse, and cultural pluralism. The inventive pragmatism of heterogeneous construction offers a path forward that bypasses the modernist dogmas of Giedion and his successors to develop new ways of building with a spirit of radical inclusion. In this light, heterogeneous construction can be seen as a methodological framework that is at once deeply historical and of critical contemporary relevance, and which has the potential to challenge our tendency to reduce construction to a series of products and received systems. The heterogeneous might instead be understood as a deep and broad body of knowledge that demands to be read, interrogated, and constantly reworked.

1 Sigfried Giedion, *Building in France, Building in Iron, Building in Ferroconcrete*, trans. James Duncan Berry (Santa Monica, CA: Getty Center for the History of Art and the Humanities, 1995), 130.

2 For further discussion, see Brett Schneider's essay, "How to Build Heterogeneously" in the present volume.

3 See Joseph Lstiburek, "The Perfect Wall," *Building Science Insights*, July 15, 2010, https://buildingscience.com/documents/insights/bsi-001-the-perfect-wall.

4 Our work in assembling this compendium was secondary by nature, and built on careful field research and documentation by many others. These sources are acknowledged in the bibliography and, in certain cases where limited documentation required reliance on a single primary source, within the case study itself. A broader range of projects from this research can be found in the See Also section of the present volume.

5 For more precise descriptions of these types and the materials that go into them, see the Glossary in the present volume.

6 Vittoria di Palma, "Radical Thought," https://www.e-flux.com/architecture/history-theory/159243/radical-thought/. We owe this reference to Ajay Manthripragada, whose design studio Distant Towers, to which we were fortuitously adjacent when teaching together at RISD Architecture, provided the initial inspiration to look outside the Western canon.

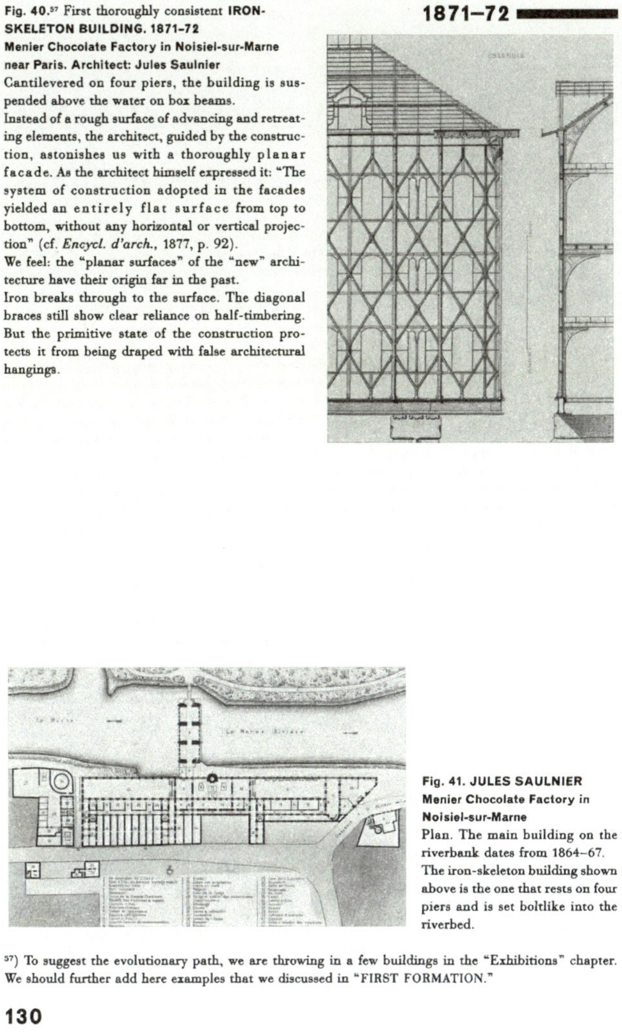
Fig. 40.[57] First thoroughly consistent **IRON-SKELETON BUILDING. 1871–72**
Menier Chocolate Factory in Noisiel-sur-Marne near Paris. Architect: Jules Saulnier
Cantilevered on four piers, the building is suspended above the water on box beams.
Instead of a rough surface of advancing and retreating elements, the architect, guided by the construction, astonishes us with a thoroughly planar facade. As the architect himself expressed it: "The system of construction adopted in the facades yielded an entirely flat surface from top to bottom, without any horizontal or vertical projection" (cf. *Encycl. d'arch.*, 1877, p. 92).
We feel: the "planar surfaces" of the "new" architecture have their origin far in the past.
Iron breaks through to the surface. The diagonal braces still show clear reliance on half-timbering. But the primitive state of the construction protects it from being draped with false architectural hangings.

1871–72

Fig. 41. JULES SAULNIER
Menier Chocolate Factory in Noisiel-sur-Marne
Plan. The main building on the riverbank dates from 1864–67. The iron-skeleton building shown above is the one that rests on four piers and is set boltlike into the riverbed.

57) To suggest the evolutionary path, we are throwing in a few buildings in the "Exhibitions" chapter. We should further add here examples that we discussed in "FIRST FORMATION."

130

Fig 1

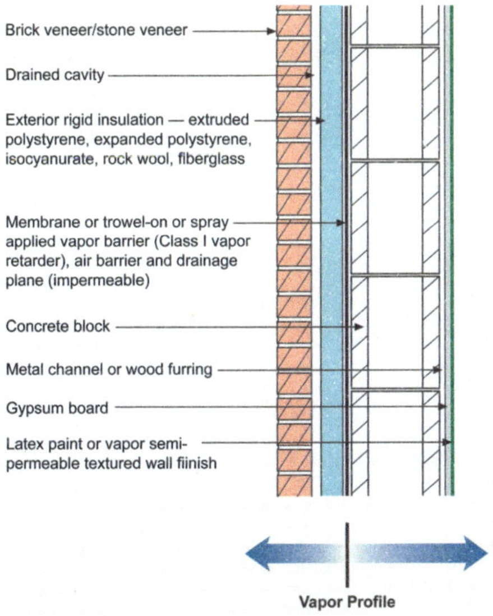

Fig 2

Fig 1 The Menier Chocolate Factory, as presented in Sigfried Giedion's *Building in France, Building in Iron, Building in Ferroconcrete.*
Fig 2 "The Perfect Wall." Diagram by Joseph Lstiburek, 2010.

The materials used in construction are at once familiar and foreign. Modern construction identifies materials that are specified within increasingly narrow bands of performance characteristics. The materials employed in heterogeneous construction are often near to their industrial equivalents but allow a wider range of techniques, performances, and effects. These broader categories are used here as both an antidote to the strictures of modern approaches and as a basis for understanding how these broader categories might begin to be combined.

By the same measure, conventional definitions of construction types, typically organized by material, strength, or fire performance characteristics, quietly augur toward the discretization of architectural materials into layers that have little to do with one another. We propose instead a relational approach to construction types that attempt to define both how materials sit relative to one another in a building or assembly, as well as how they perform.

The drawings on subsequent pages are categorized and color-coded according to this schema to allow the reader a quick and intuitive grasp of how materials operate within each example without sacrificing detail or clarity.

Constructions

Collages	The combination, insertion, or juxtaposition of two or more unlike materials.
Fills	Building elements with two or more materials in the same plane.
Sandwiches	Buildings with linear or wall-like elements in which outer faces bound a filling material.
Scaffolds	Buildings in which the process of construction, maintenance, or inhabitation is made explicit.
Stacks	Buildings or elements that layer or stack materials vertically.
Wraps	Buildings or elements where materials are bound together in concentric layers.

Ceramic	Fired or baked earth-based materials. Brick, Terra-cotta
Concrete	Mixed and unmixed cement-based materials. Cement, Concrete, Lime, Ash
Earth	Unfired loam, clay, humus, and other soils. Sand, Silt, Clay, Loam, Humus, Mud
Metal	Elements or compounds that generally exhibit the structural properties of malleability and ductility. Iron, Steel, Aluminum, Brass, Bronze, Copper
Other	Synthetic polymers, glasses, and animal products. Glass, Plastic, Bone, Hide

Plants	Non-woody natural materials whose fibers may form all or part of a structural component. May be used independently, woven into a textile, or as reinforcement for earth and ceramic matrices. Grass, Bark, Leaves, Palm shoots
Stone	Hard, solid, nonmetallic mineral culled from the earth as a material suitable for a wide range of structural purposes.
Textile	Natural or synthetic fibers that are woven, spun, or knitted to create a cord or fabric. Canvas, Felt, Rope
Wood	Dried trunks, branches, and other parts of trees and bushes whose fibers and lignin work as a composite in resisting bending, compression, and tension. Timber, Bamboo, Engineered Wood

Examples

Name	ACHUM SHRINE
Location	Bafut, Cameroon
Year	1895 (reconstruction)
Construction	Wood and Plants as Wrap

The Achum Shrine is the core of the sacred Fon's (king's) palace in Bafut. It houses a devotional object and is accessible only by the king. Originally built 400 years ago, it was reconstructed in 1895, and is representative of certain local construction techniques. Set in a square on a tall stone plinth, carved timber columns support a pyramidal thatched roof framed in timber, with rafters extending radially out from the roof's peak to the perimeter. A multilayered wall set back from the peristyle is composed of bamboo pillars tightly bound together using grass lashing, and set in front of a grass mat that provides a tighter enclosure. The tight spacing of the lashed bamboo wall mirrors the vertical stonework of the plinth and provides additional support for the ceiling and roof as well as lateral stability for the structure as a whole. The intricate nature of the wall and roof construction stands in complement to the decorative carvings that mark the perimeter columns of the building.

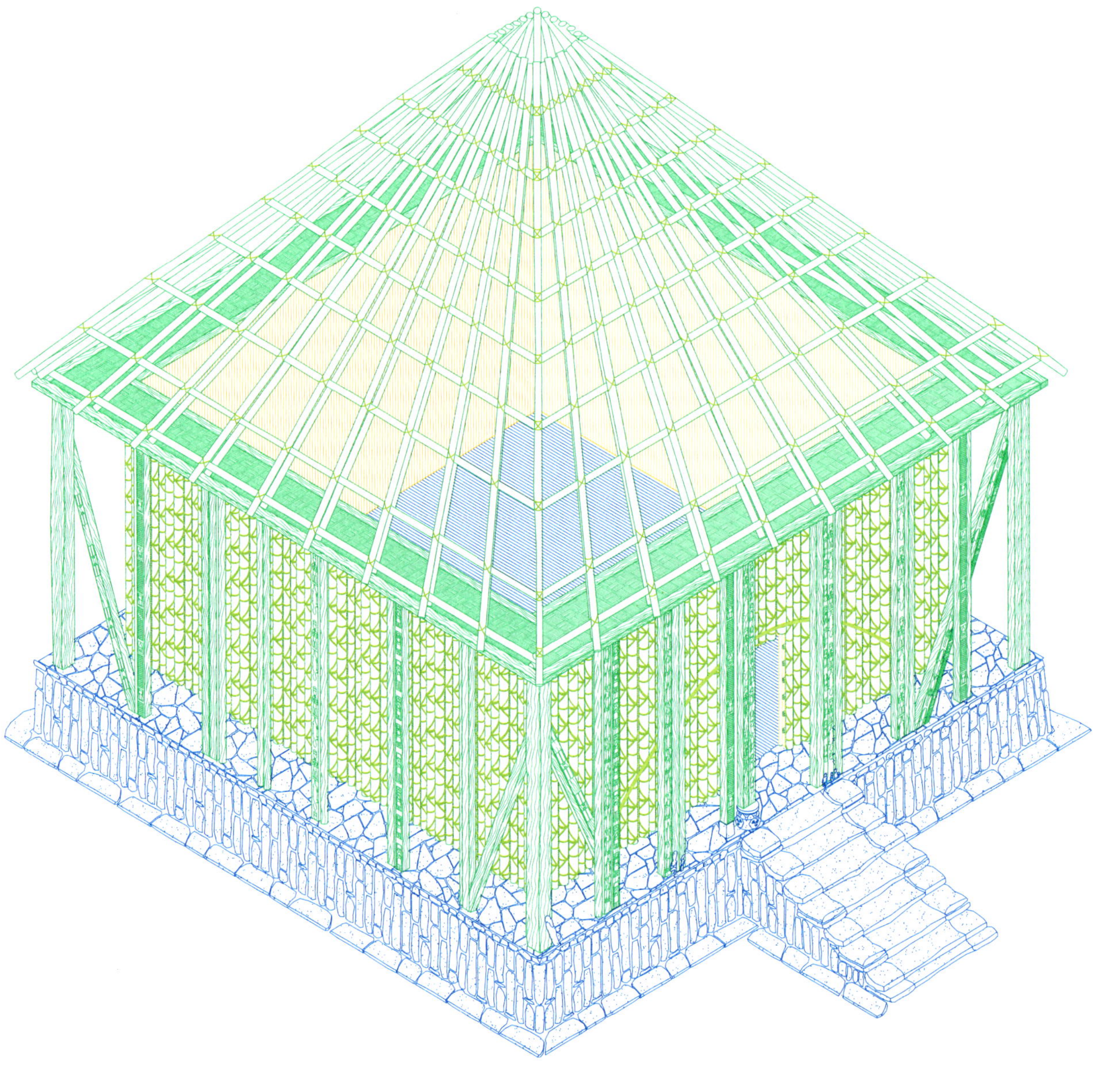

Name	BLACK TENT
Location	Tibet
Year	Unknown
Construction	Textile and Wood as Scaffold

Based on fieldwork by Gordon Clarke and the Institute for Nomadic Architecture.

The black tent originated during the Ancient Mesopotamian era, and is a nomadic tensile structure made primarily of yak hair and wood posts. The roof textile is carried by two compressive wood posts that support the main horizontal ridgepole, and together with smaller vertical props located outside of the tent, stay a network of ropes over which the fabric enclosure drapes. As with modern tensegrity structures, the individual components cannot stand on their own but rely on one another for stability.

The textile is made of spun and woven yak hair, which is strong and naturally water-resistant. It is woven in strips which are sewn together into two halves that are then joined at the center of the tent above the ridgepole with loops and yak horn toggles. An opening in the roof allows light to enter and smoke to escape. The black tent evolved with the nomadic lifestyle of its inhabitants. As a result, the yak provides the materials and transportation for the tent in addition to providing sustenance for its inhabitants.

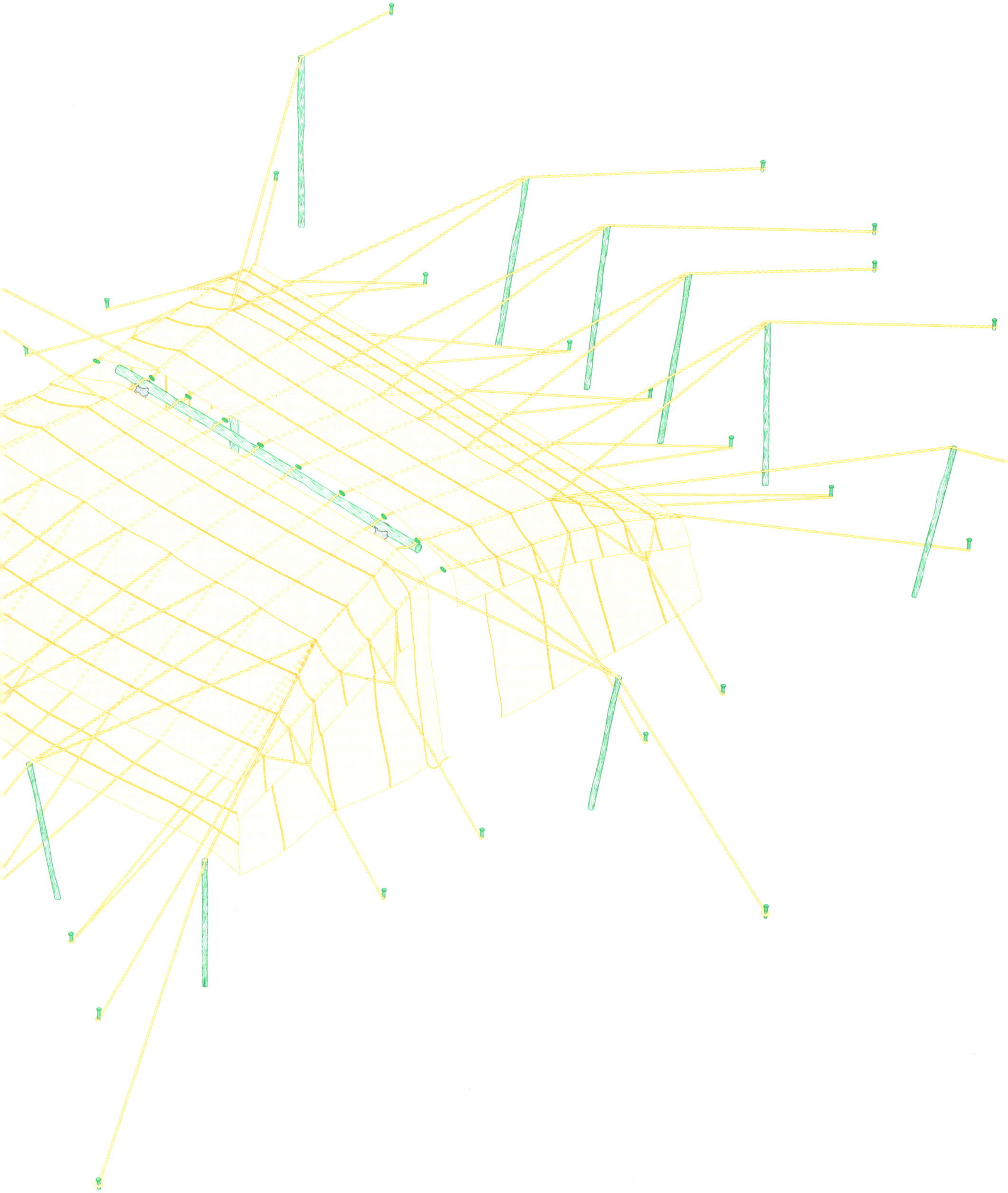

Textile, Wood, Other

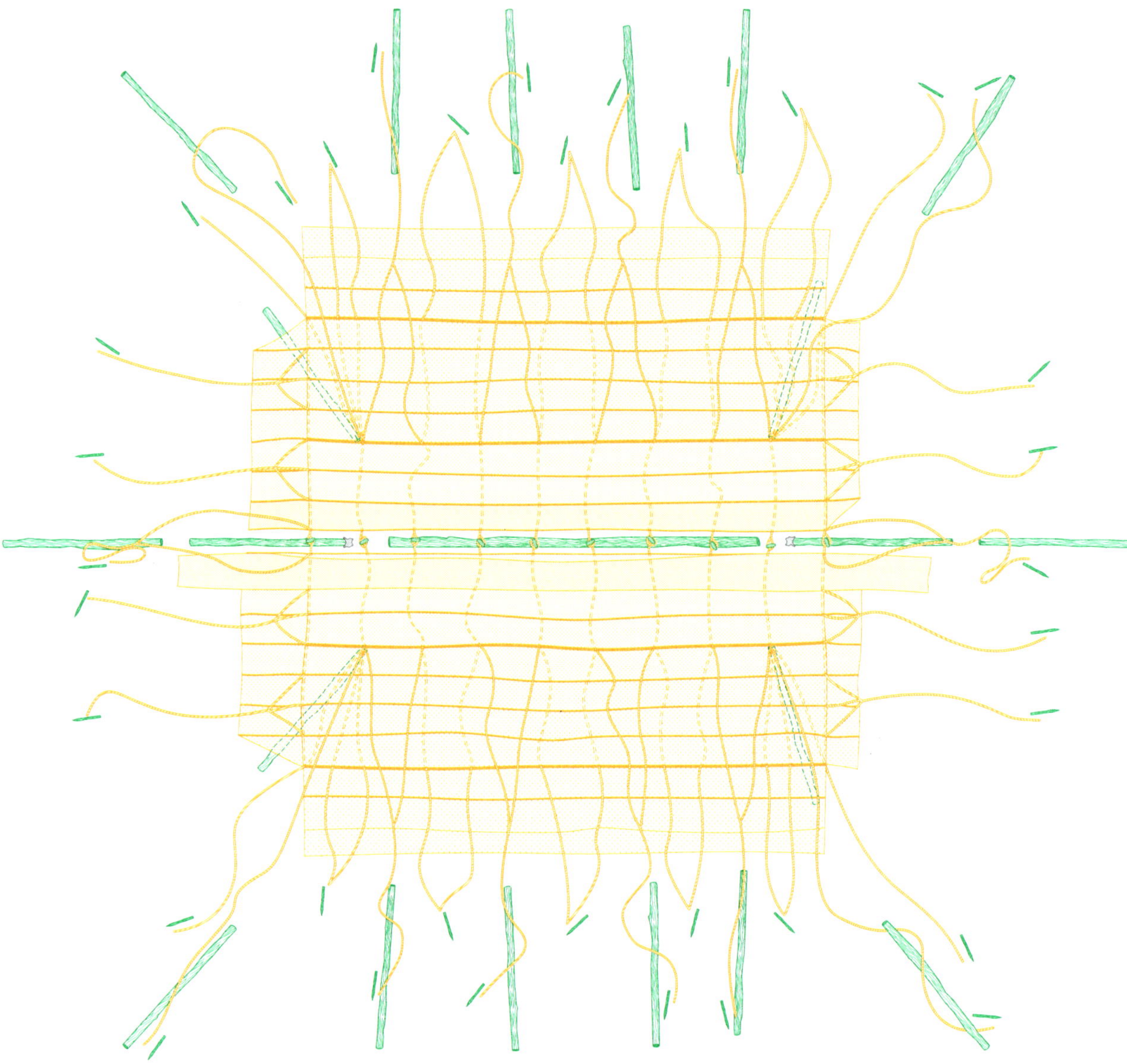

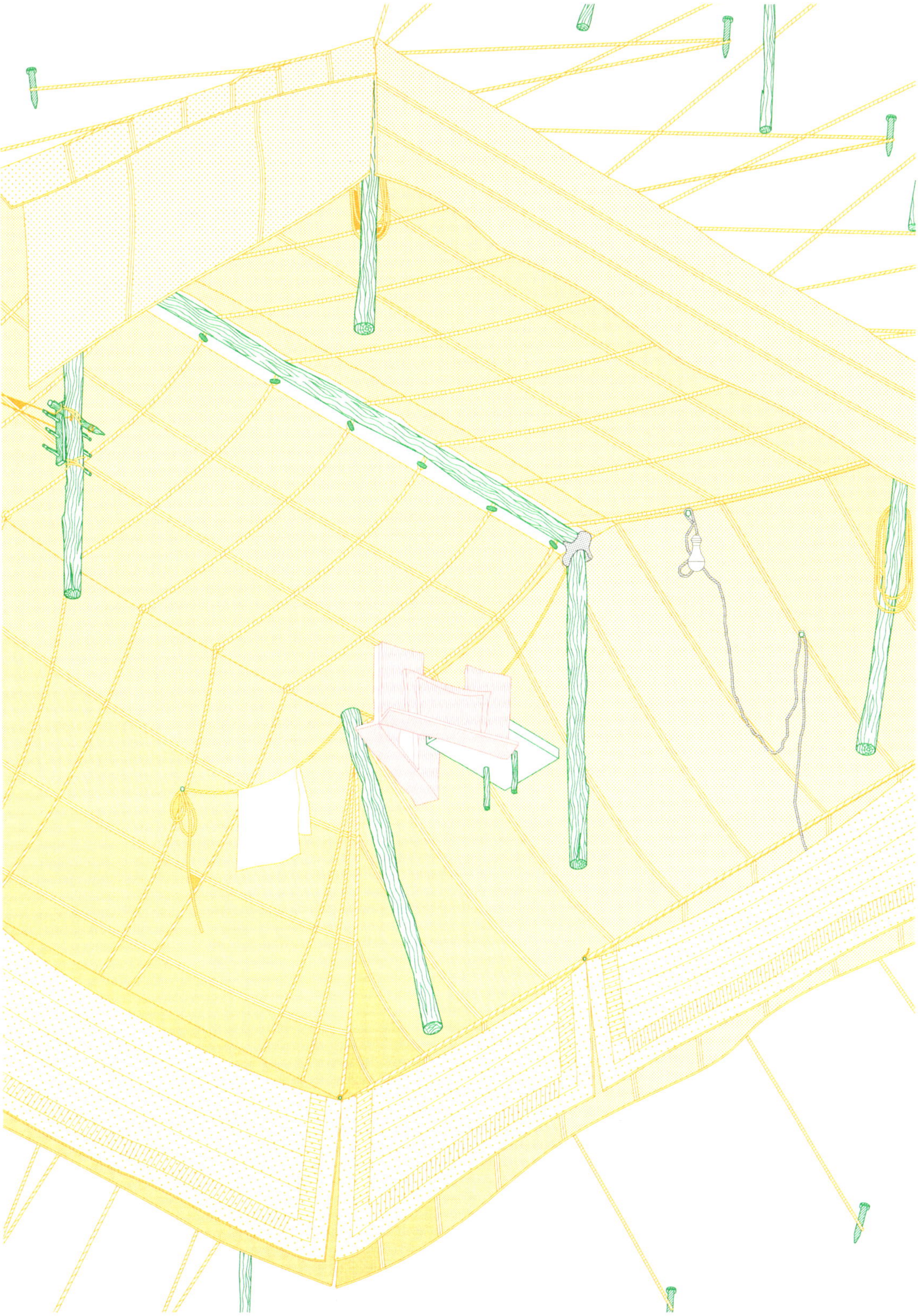

Name	CATTLE SHED IN COSSIAT
Location	Ferme de Cossiat, France
Year	ca. 1844
Construction	Earth and Ceramic and Stone and Wood as Scaffold

Based in part on research from Roger Boltshauser, *Pisé–Rammed Earth Tradition and Potential* (2019).

The cattle shed is part of a larger farm ensemble that is representative of the revival of earth building in the region following the influence of architect François Cointereaux, who systematized the building method during the early Industrial Revolution. The building interleaves four different materials—earth, brick, stone, and wood—in a sophisticated manner based on relative availability and sturdiness. The body of the wall is made of compressed loam gathered from the site and makes use of standardized formwork, which leaves voids in the wall once removed. Horizontal layers of lime are laid with each lift as an erosion control technique. Stone masonry extends from the foundation up the base of the wall and transitions to serrated brick coursing at the building corners to reinforce the walls during compaction and provide further protection from erosion. The extended eaves of the timber and clay-tile roof gain support from buttresses that tie back to the formwork voids referenced above.

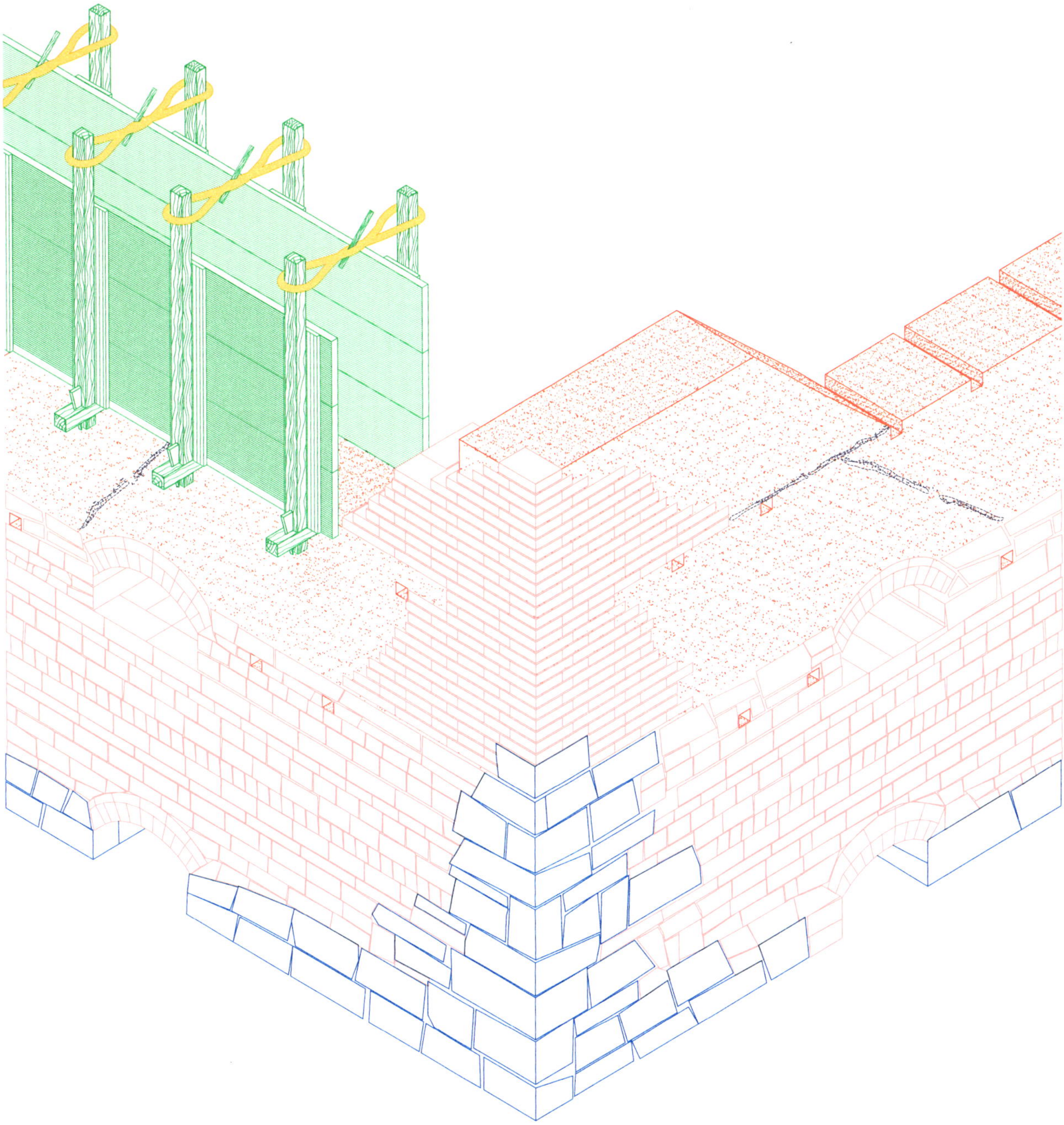

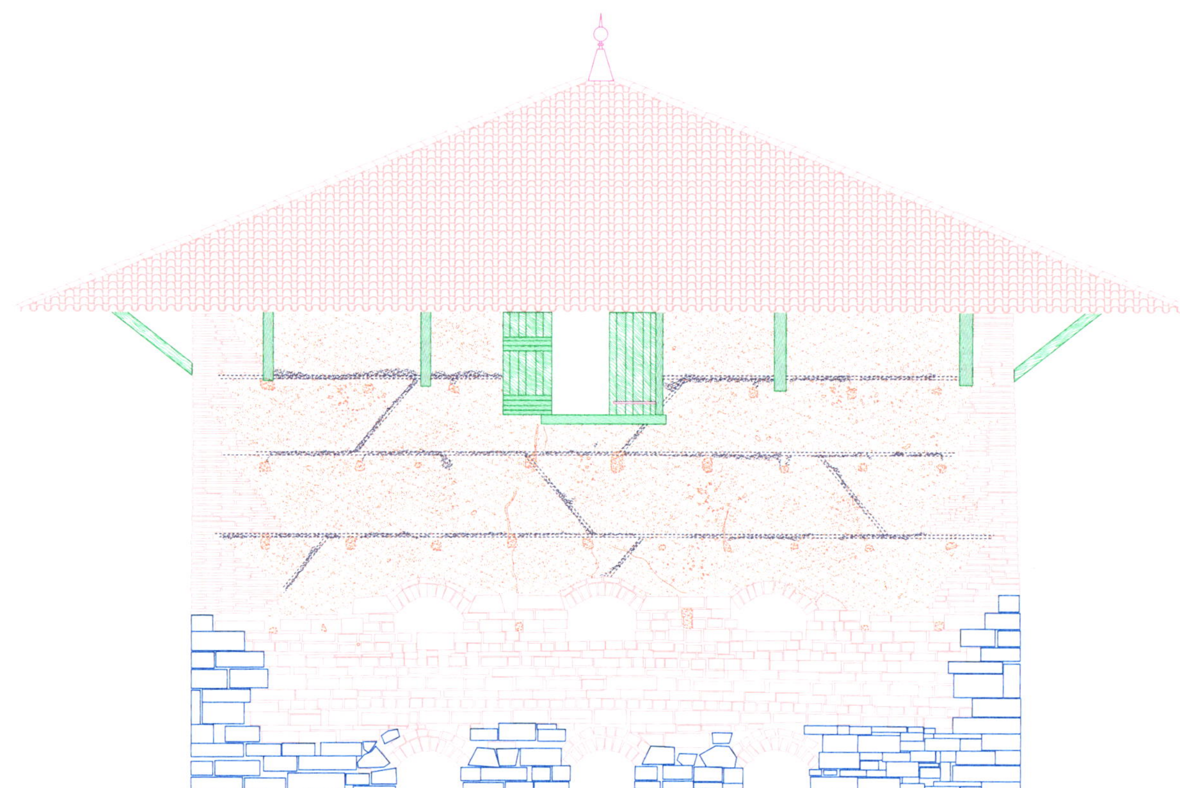

Earth, Ceramic, Stone,
Wood, Concrete

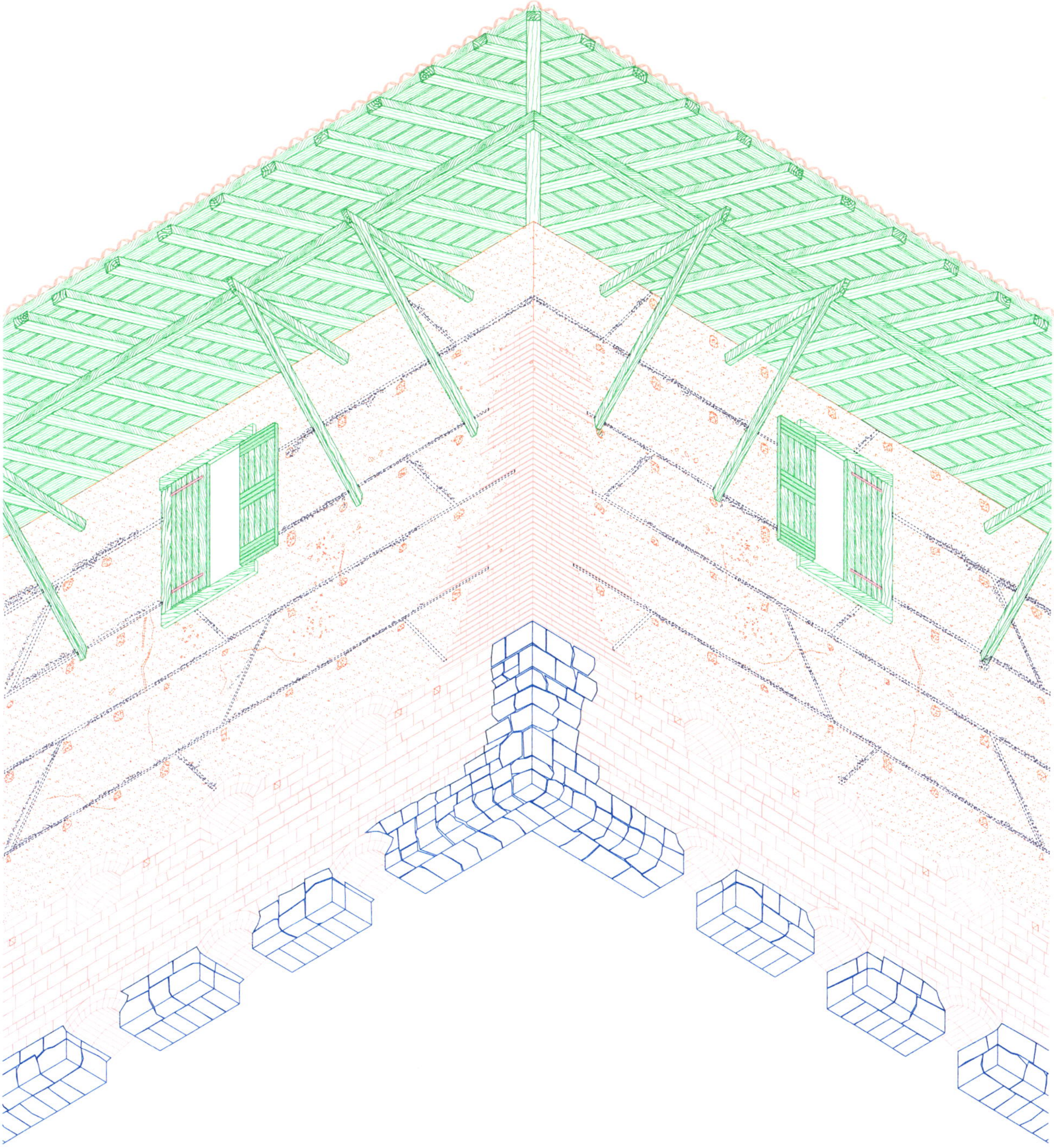

Name	GÖZ DOLMA
Location	Black Sea Region, Turkey
Year	19th century
Construction	Wood and Stone as Fill

Göz dolma is one of several types of vernacular house construction from Anatolia that use composite timber frame and masonry infill walls (*hımış*). The perimeter walls of the building are formed from a tightly gridded frame of wood members at approximately 20-25 cm spacing with a single shaped river stone of consistent size in each cell of the grid to form an approximately 10 cm thick wall. A lime-earth mortar is then filled around each stone to seal the wall. The type usually includes stone foundations and wood framed hip roofs. The composite wall system in göz dolma construction provides a highly redundant frame with natural damping characteristics that results in good seismic performance. The interior partitions are a type of stud-frame construction with horizontal boards fastened between wood posts to form a cavity wall. The walls are plastered or faced in intricately carved wood paneling.

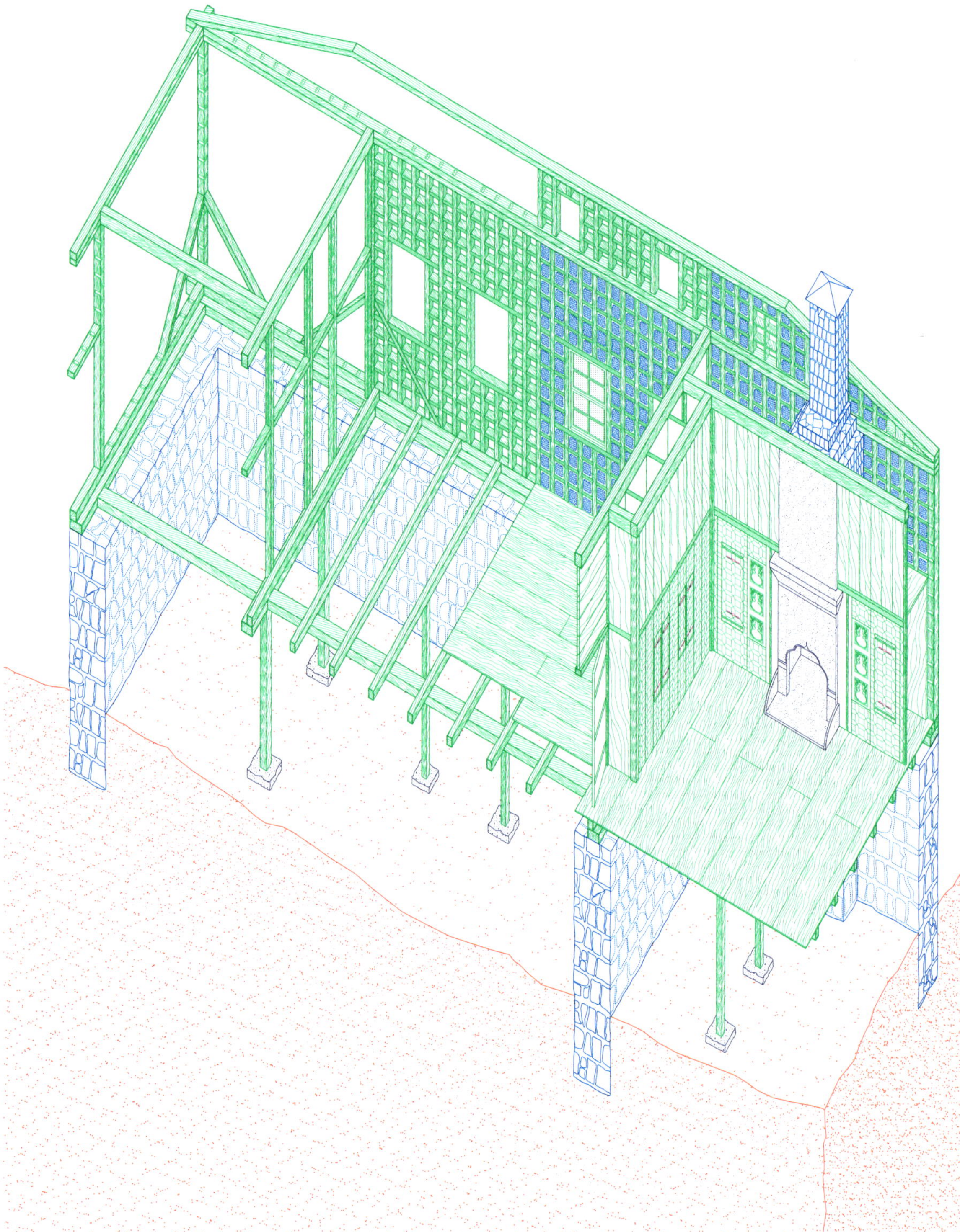

Name	GREAT MOSQUE OF DJENNÉ
Location	Djenné, Mali
Year	1907 (reconstruction)
Construction	Earth and Wood as Scaffold

This large-scale earth construction has roots dating back to the thirteenth century, but the current building is a 1907 reconstruction by Ismaila Traoré. The mosque is built from cylindrical sunbaked earth bricks typical of the region; sand and earth plaster; and bundles of rônier palm timbers. The sticks protrude from the perimeter walls on both sides, supporting both interior roof framing and forming a scaffold for members of the congregation to climb during the annual maintenance festival, when the facades are replastered with fermented mud. The palm sticks form the expressive facade of the building and may help reduce cracking of the earth structure. The mosque's interior is characterized by an earth floor and tightly spaced massive earth pillars with pointed arches that support a palm branch roof. This, in turn, is surfaced with mud waterproofing and has ventilation shafts shielded by removable clay pots.

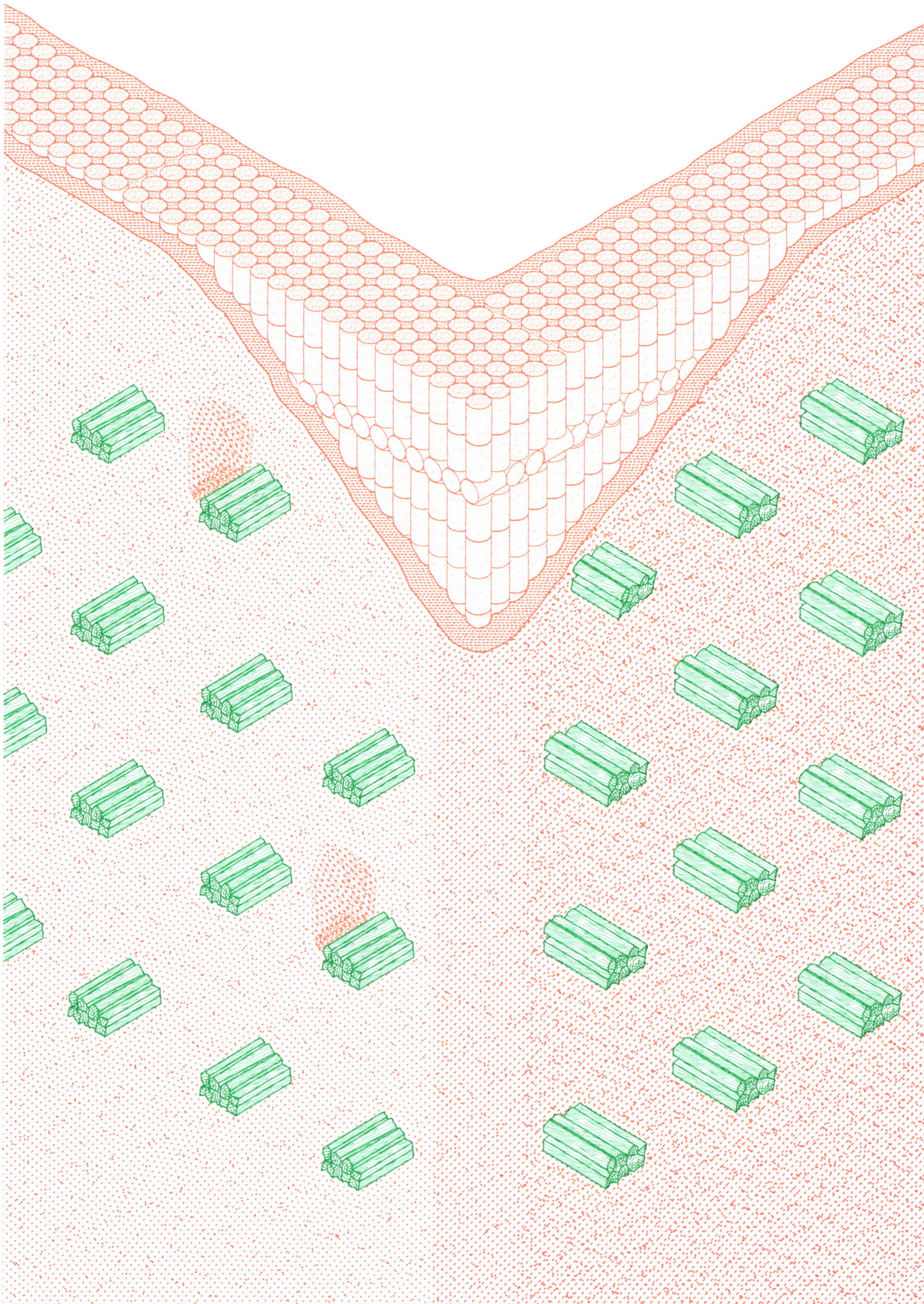

Name	KATH-KHUNI
Location	Himachal Pradesh, India
Year	Since 18th Century
Construction	Stone and Wood as Stack

Based on fieldwork by teams led by
Mansi Shah and Jay Thakkar.

Kath-khuni is a vernacular construction type from northern India composed of stacked stone and wood walls. The articulation of the offset courses of wood at the corners is the focus of the decorative style of the type. The construction of the perimeter walls consists of a double-skin of alternate courses of dry stone masonry and local, rot-resistant Deodar cedar wood. The wood and stone courses include alternating wooden beams and dressed stone lines laid parallel to each other with a gap in-between filled with loose stone. The wood elements are tied with cross braces or dovetailed *maanvi* and wood pins for vertical connection. The wall is finished with mud plaster on the interior to seal the wall and improve its insulating ability. The resulting construction is naturally earthquake-resistant due to the interaction between the relatively ductile wood frame and damping of the dry laid stone.

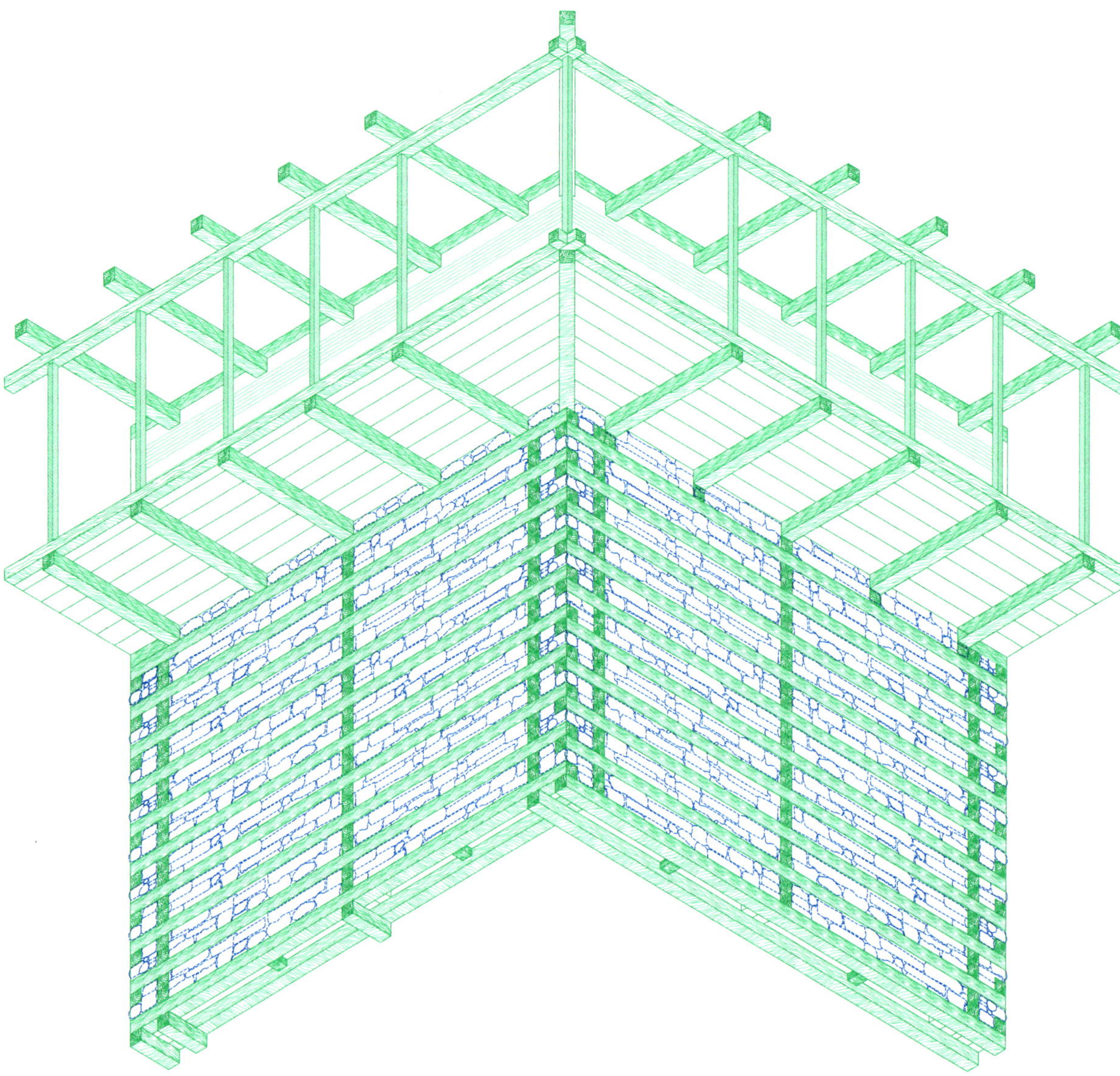

 Stone, Wood

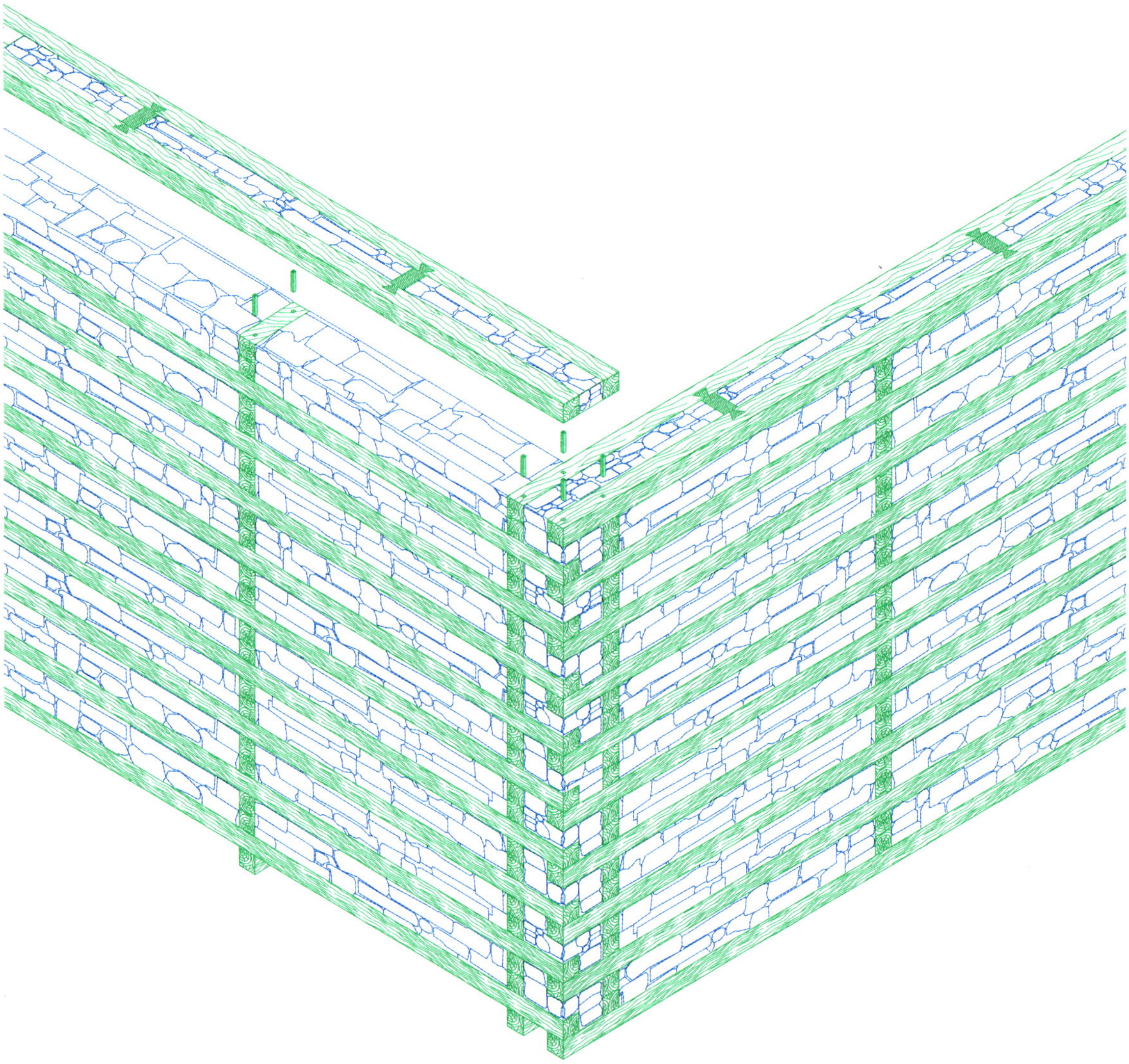

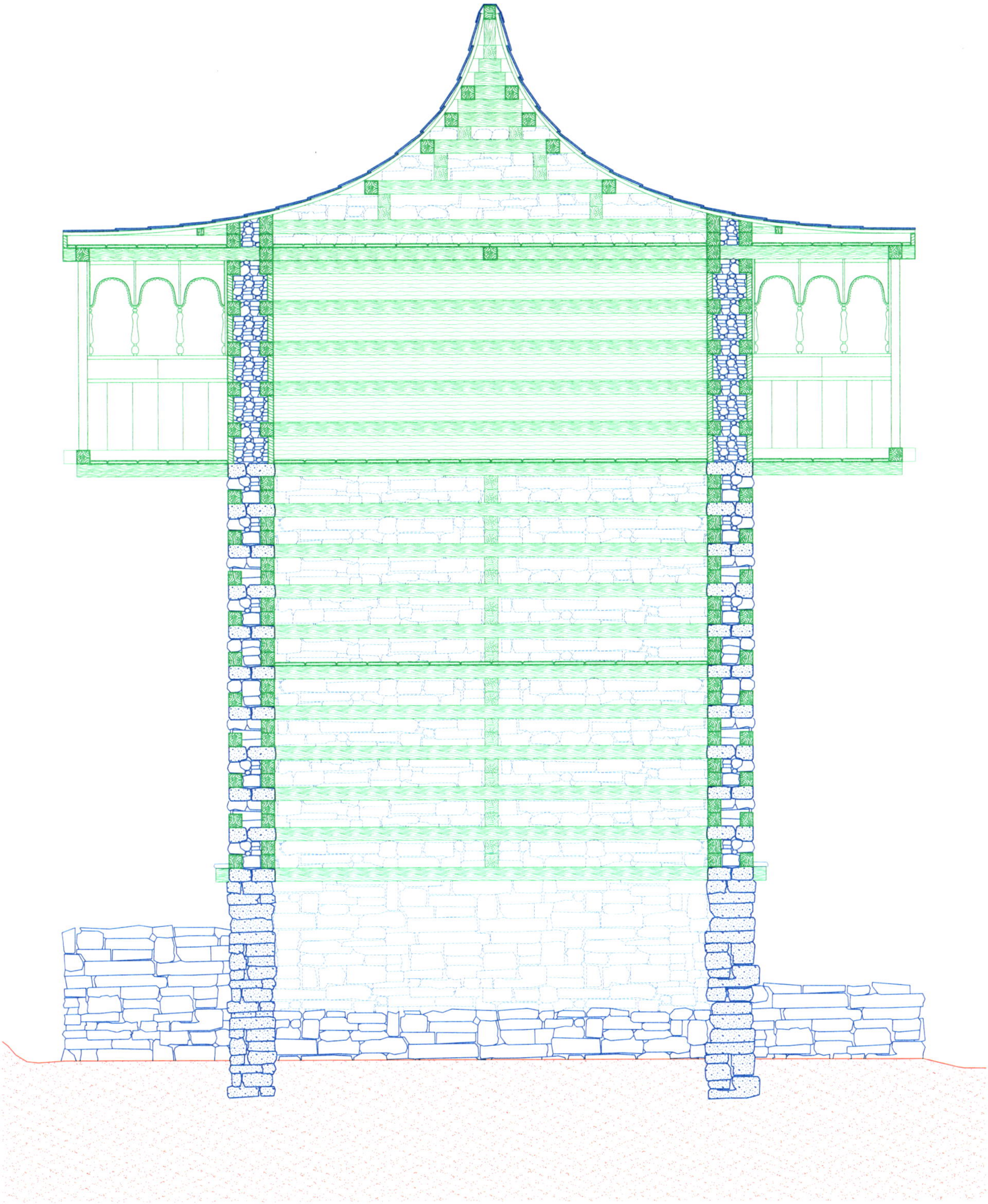

Name	KILN TOWER FOR THE BRICKWORKS MUSEUM
Location	Cham, Switzerland
Year	2021
Construction	Earth and Wood and Metal as Stack

The Kiln Tower is a kiln and exhibition building designed by Roger Boltshauser and students at TU Munich. The building makes use of prefabricated earth blocks compressed by a heavy timber roof tied with steel cables to the concrete foundation. In the context of a tower-shaped building, the post-tensioning locks the inherently irregular blocks into an effective whole as a series of springs in tension and compression. There are planks of cross-laminated timber under each prefabricated block to facilitate lifting and transportation. The planks also form the horizontal joints between blocks to remove additional on-site labor, accentuate the modular construction of the blocks, and provide attachment for the outboard wood weather drip. The resulting components remain discrete as elements so that the cables can be loosened and the entire structure easily disassembled for recycling or reuse as necessary in the future.

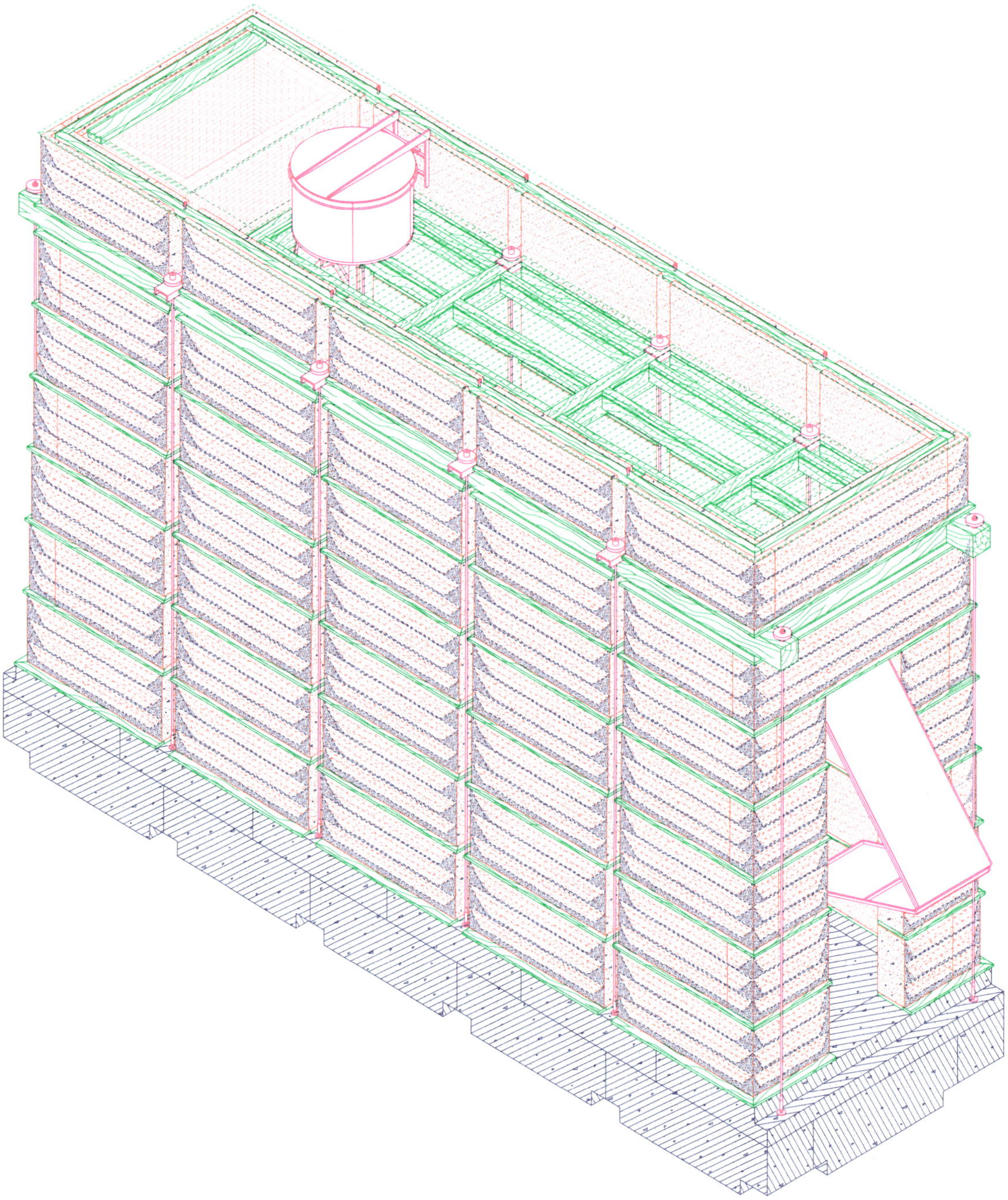

Earth, Wood, Metal,
Concrete, Ceramic

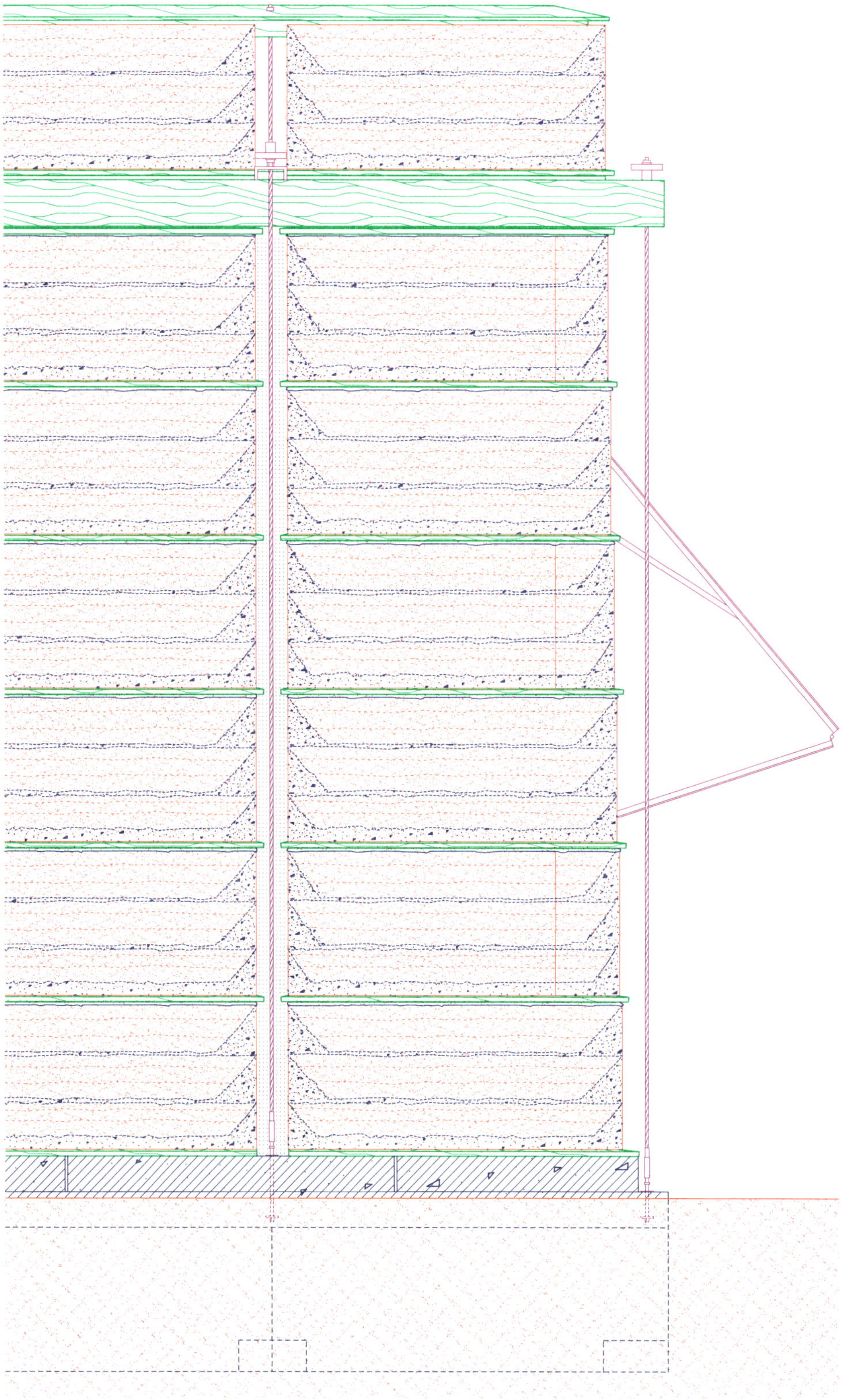

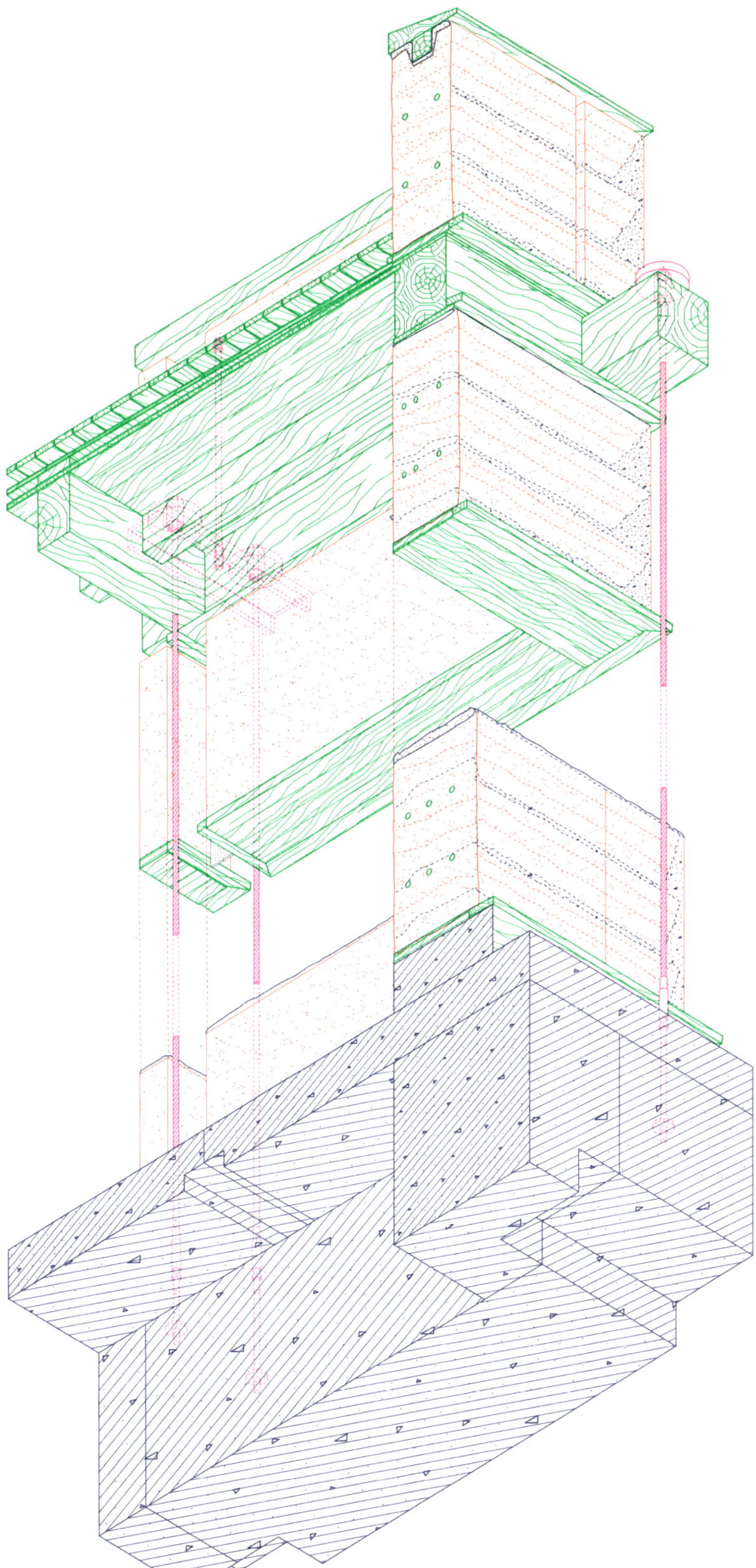

Name	MENIER CHOCOLATE FACTORY
Location	Noisiel, France
Year	1872
Construction	Metal and Ceramic as Fill

Designed by architect Jules Saulnier and structural engineer Armand Moisant, the building bridges over the Marne River with iron beams bearing on a set of existing stone piers. The four structural walls of the factory are constructed of a diagonal iron lattice infilled with a double layer of hollow brickwork. The brick lends mass—and likely added stiffness—to a frame that is both unusually light and imperfectly triangulated. The brick is plastered on the interior and glazed on the exterior to produce a vibrant polychrome facade. Inside, shallow brick jack arches span between iron beams to create composite floors that brace the walls. The first and second floors are supported by beams on iron columns, while the third is suspended from the roof trusses above, leaving the second floor free of supports. The iron frame coupled with the flatness of the facade made the building a model of iron construction for early modernist architects, who downplayed the decorative qualities of the surface and the role of the infill brick.

Metal, Ceramic

USINE HYDRAULIQUE DE NOISIEL

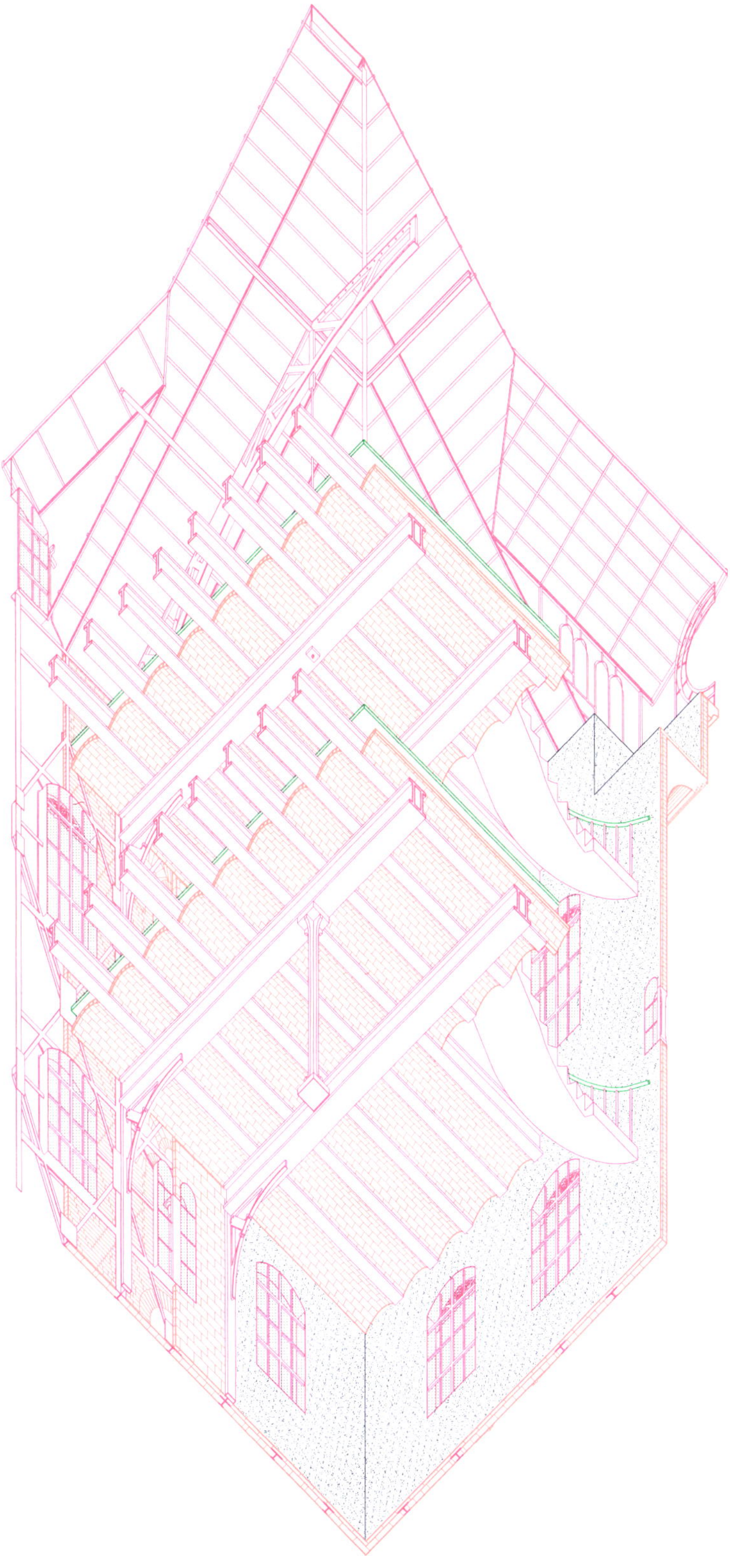

Name	NATIONAL MUSEUM OF ROMAN ART
Location	Mérida, Spain
Year	1986
Construction	Ceramic and Concrete as Sandwich

This archaeological museum designed by Rafael Moneo is built over the ruins of the Roman colony of Augusta Emerita in Mérida, Spain, using a contemporary translation of a construction technique derived from the ruins themselves. The main exhibition space is structured by a series of tall, parallel walls made of concrete-infilled brick to invoke and update the Roman *opus testaceum* (brick-faced concrete). The walls, columns, and arches were laid up in the shape of hollow tubes made of hard-fired bricks with deeply recessed mortar joints to create a more abstract, planar brick surface. These tubes were then filled incrementally with reinforced concrete up to the ground floor, and unreinforced concrete above (there is a small amount of reinforcement for continuity). Together with the concrete fill, the slender brick tubes form a load-bearing structure that supports a series of reinforced concrete floor slabs and glazed, precast concrete roof lanterns. In this case, the formwork for concrete casting is reimagined as permanent, to produce both the final finish for the walls and a structural ally for the unreinforced concrete.

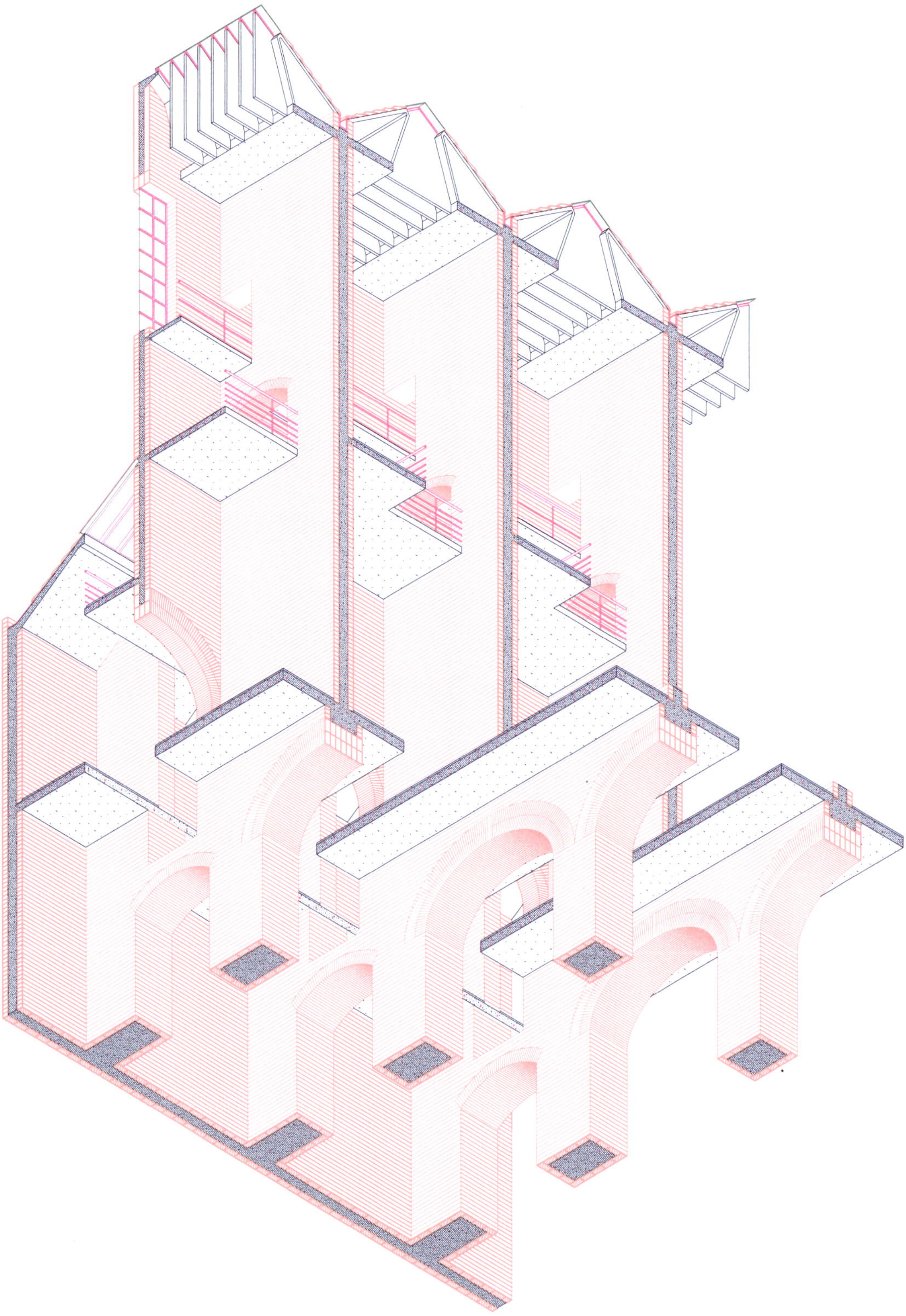

Ceramic, Concrete, Metal

Ceramic, Concrete, Metal

Name	SCHOOL IN ORSONNENS
Location	Orsonnens, Switzerland
Year	2017
Construction	Wood and Concrete and Metal as Collage

This school by TEd'A Arquitectes is a contemporary reinterpretation of traditional barns in the region. While taking inspiration from the barns' overall stone base and wood frame construction, the interior reveals a more complex structural interplay between a mass timber frame and exposed concrete. Timber and concrete are employed both as independent and interwoven systems. The classrooms that form the bulk of the program are supported using an innovative timber-concrete composite floor system, with closely spaced timber beams working in tandem with the slab above via shear studs cast into the concrete, increasing the span of both systems while reducing the carbon-intensity of the whole. In the central atrium, the two systems separate, with concrete slabs cantilevering out into the space around an expressive timber column that extends from the ground upward, supporting the peak of the roof while branching out to meet bracing points around the perimeter of the space.

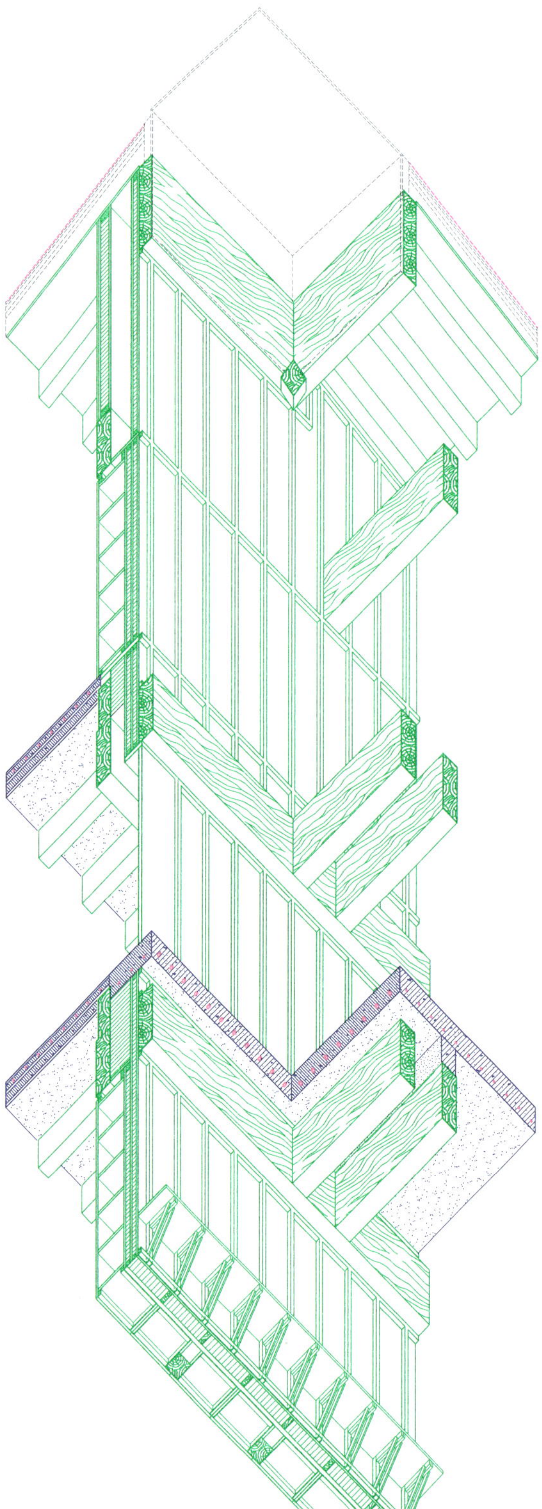

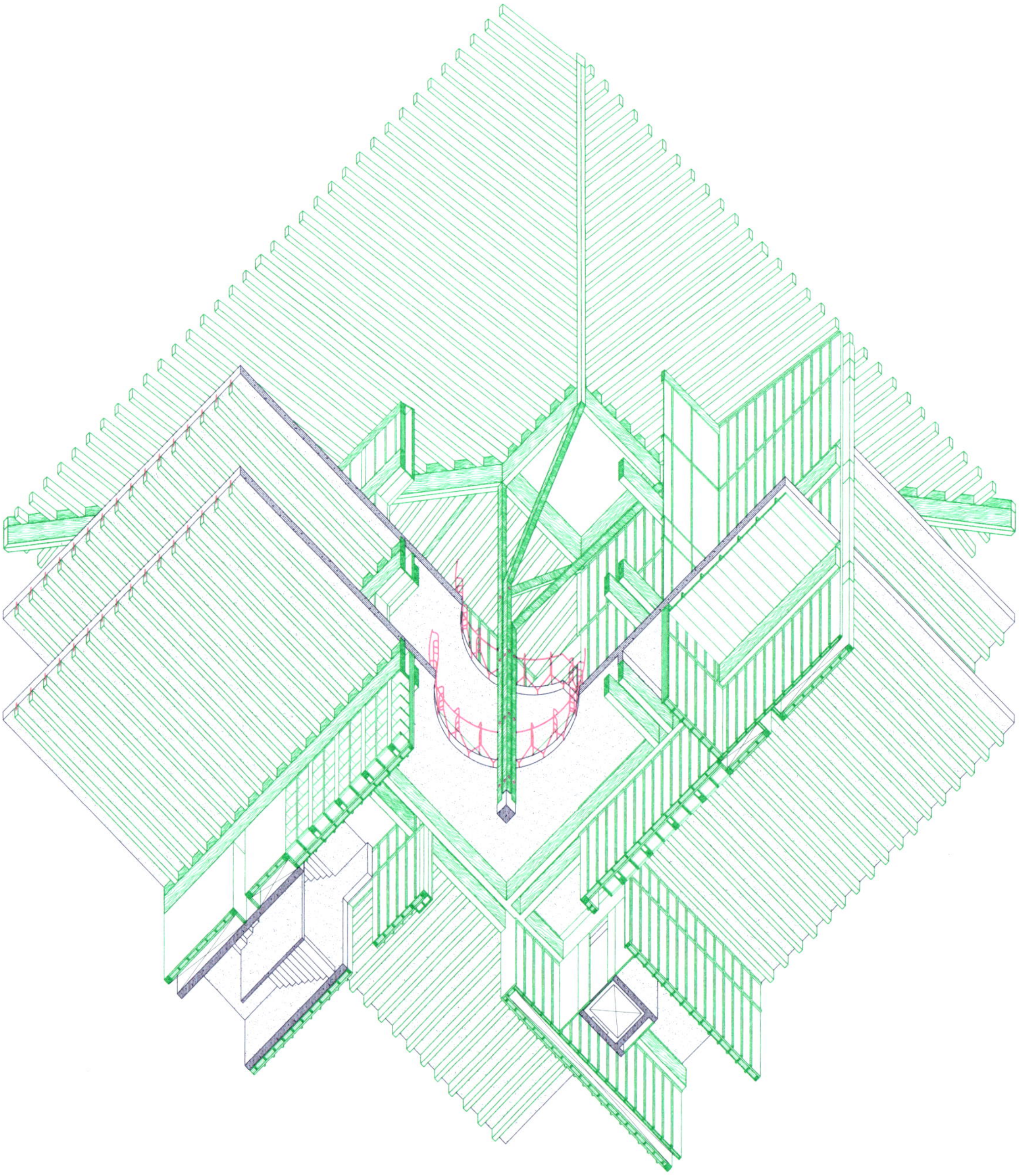

Name	SENECA BARK LONGHOUSE
Location	Victor, New York
Year	1998 (reconstruction)
Author	Seneca Tribe
Construction	Wood and Plants as Sandwich

This reconstructed building is an example of a once-common construction technique among Indigenous tribes of the region in what is now upstate New York. Typical longhouses, which were used as communal dwelling spaces, measured approximately six meters wide and up to sixty meters long depending on the number of families living within. The walls of this example are made of elm bark cladding panels sandwiched between an interior wooden elm frame and smaller exterior timber members lashed together with rope. The wood frame is doubled on the interior to form an outer structural bay holding two-tiered bunks on each side for sleeping and storage. While all of the components for the structure are derived from elm trees—posts, bark strips and bark fiber rope—each operates in a clearly distinct way to bind together the whole of the structure and enclosure.

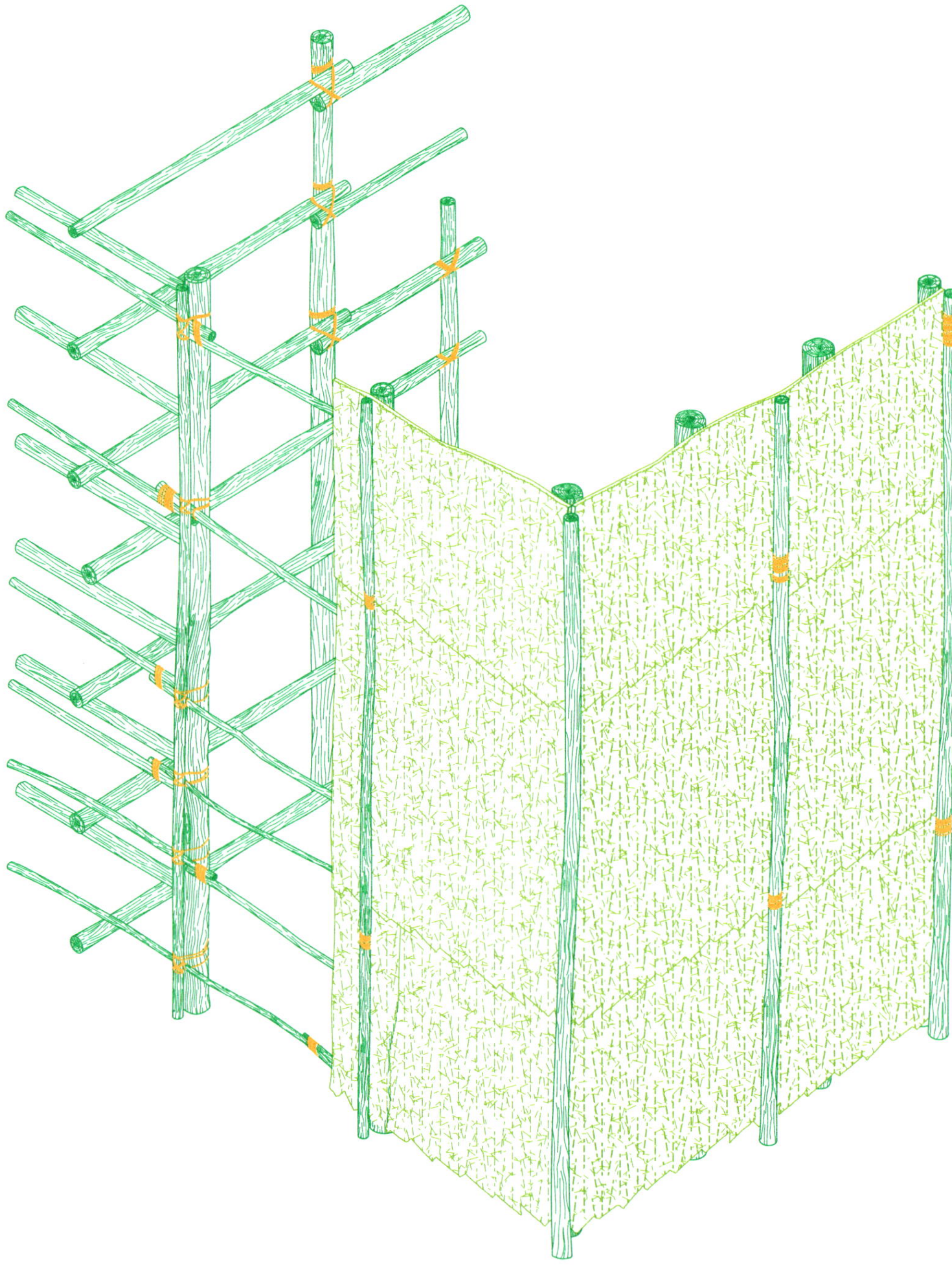

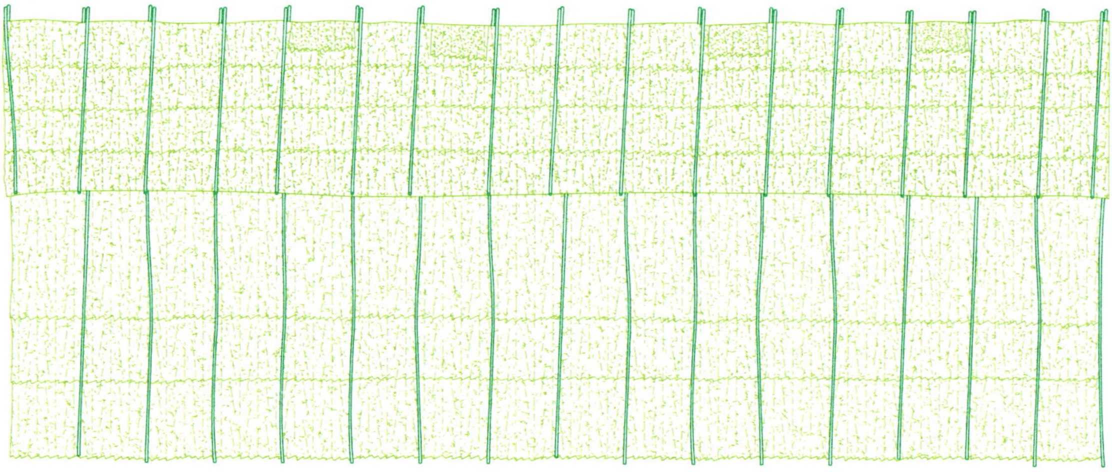

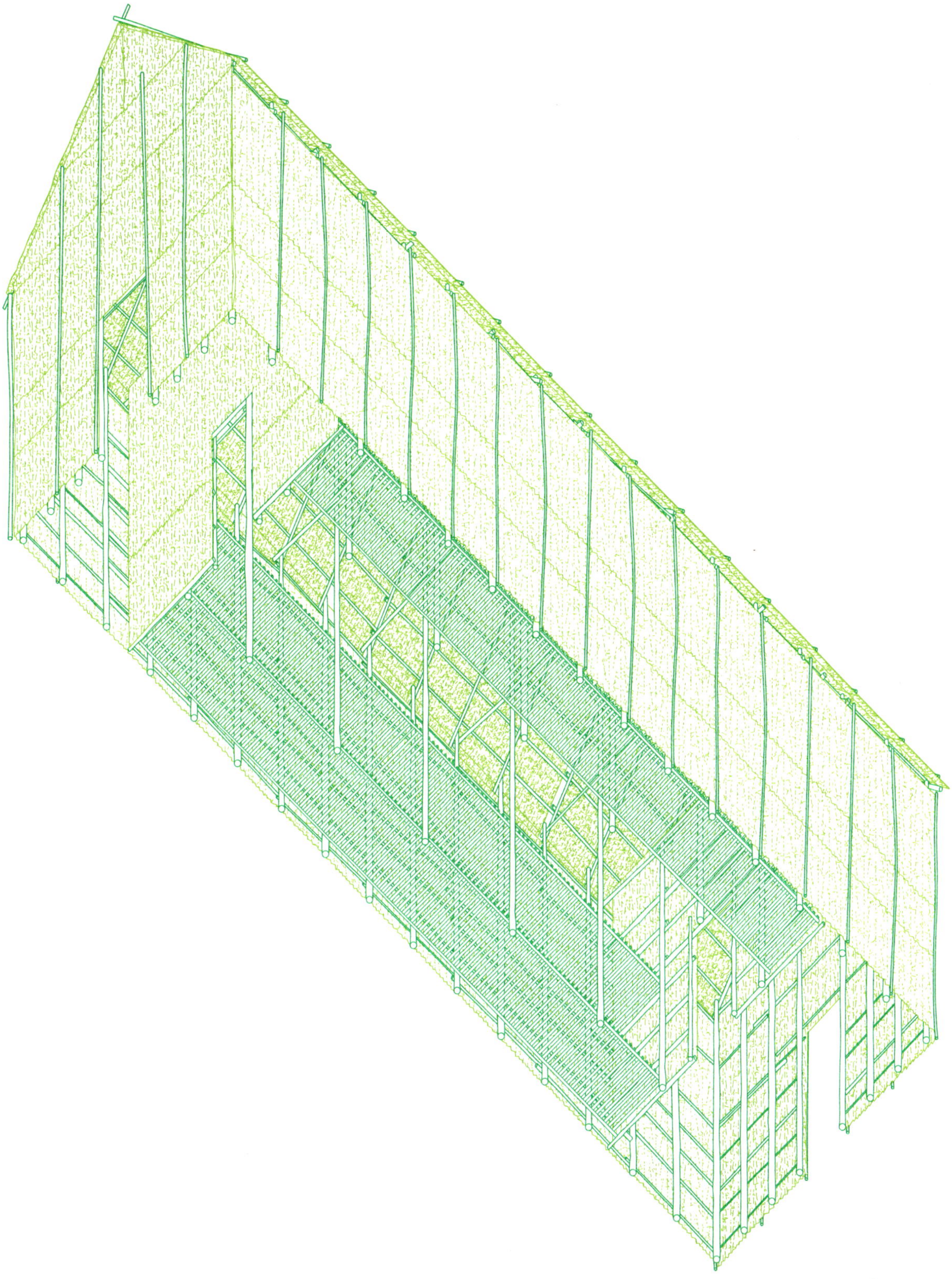

Name	TOMIOKA SILK MILL
Location	Tomioka, Japan
Year	1872
Construction	Wood and Ceramic as Fill

This Meiji era building was designed by French engineer Edmund Auguste Bastien as part of the national push to industrialize Japan. The building included a lower level for industrial reels for silk production and upper levels for warehousing of silkworm cocoons. The heavy timber frame of the building is infilled with brick at the perimeter, an unusual combination for a building of this size. The brick infill includes large openings for windows and arches in places for support without the wood frame. Brick was uncommon to Japan at the time, requiring a kiln to be built to produce the supply of brick for the mill's construction. Like many timber frame and masonry infill structures, this hybrid frame of wood and brick has performed well in this high-seismic environment.

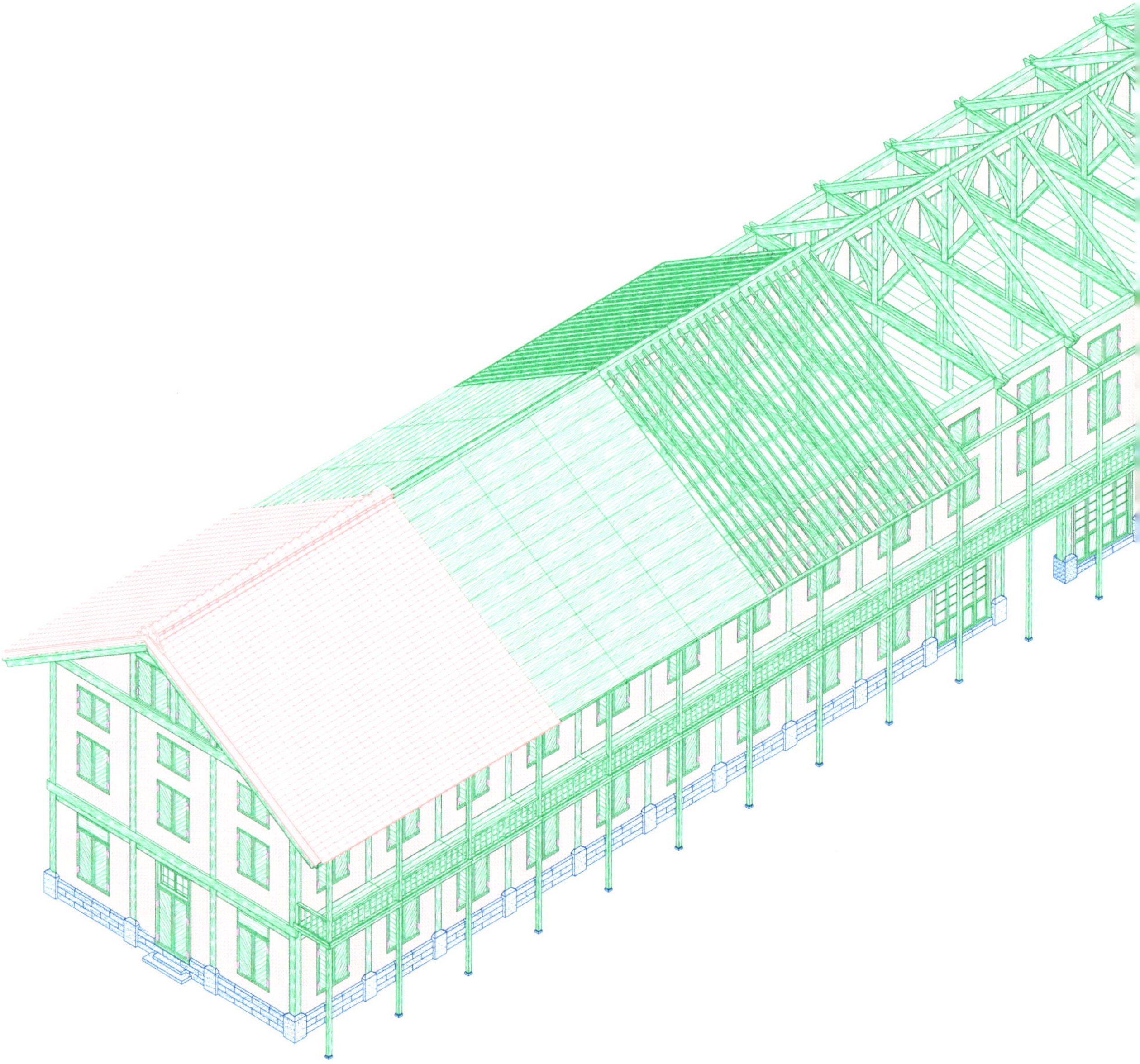

 Wood, Ceramic, Stone

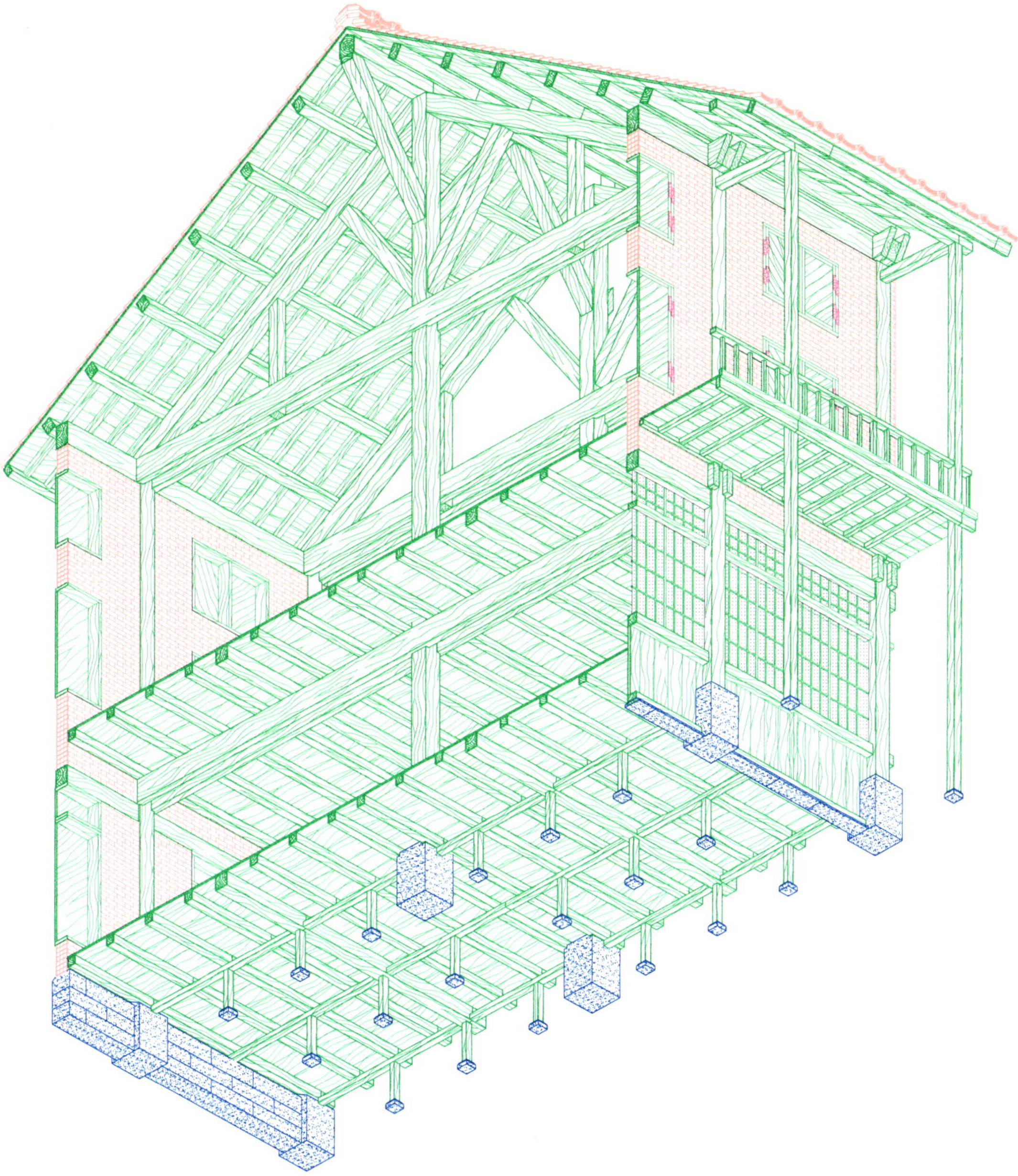

Name	TULOU
Location	Fujian, China
Year	Since 12th century
Construction	Earth and Wood as Wrap

Hakka Tulou are large collective dwellings constructed in southeastern Fujian that can each house up to eight hundred people. Typically round or square in plan, they are characterized by a defensive rammed earth perimeter wall that shelters the living units, shared spaces, and a central courtyard on its interior. The thick rammed earth wall is inclined inward on the outside and its interior face steps in as it rises. The wall is protected from water damage by a high stone foundation and an overhanging eave topped with clay roof tiles. The earth wall is rammed in blocks with movable formwork; bamboo branches inside the wall act as tensile reinforcement and tie the blocks together. A post-and-beam timber structure leans against the massive bounding wall and is supported on the steps of the wall face. The wall supports a radially organized timber frame and roof structure that houses three to four stories of living units. The earth wall is typically blank and often has only one gate and small window openings on the exterior, contrasting with the intricately decorated, airy timber structure on the interior.

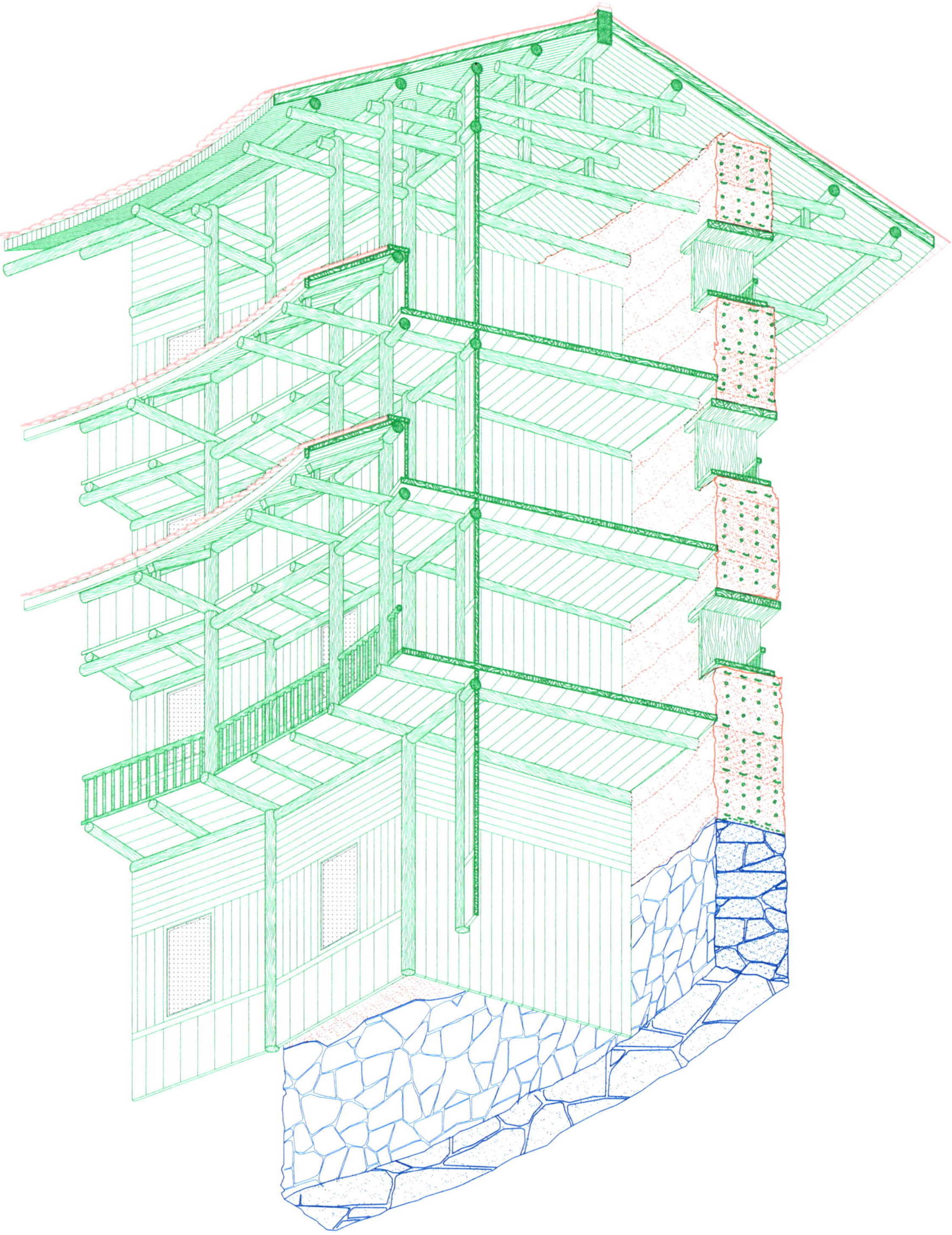

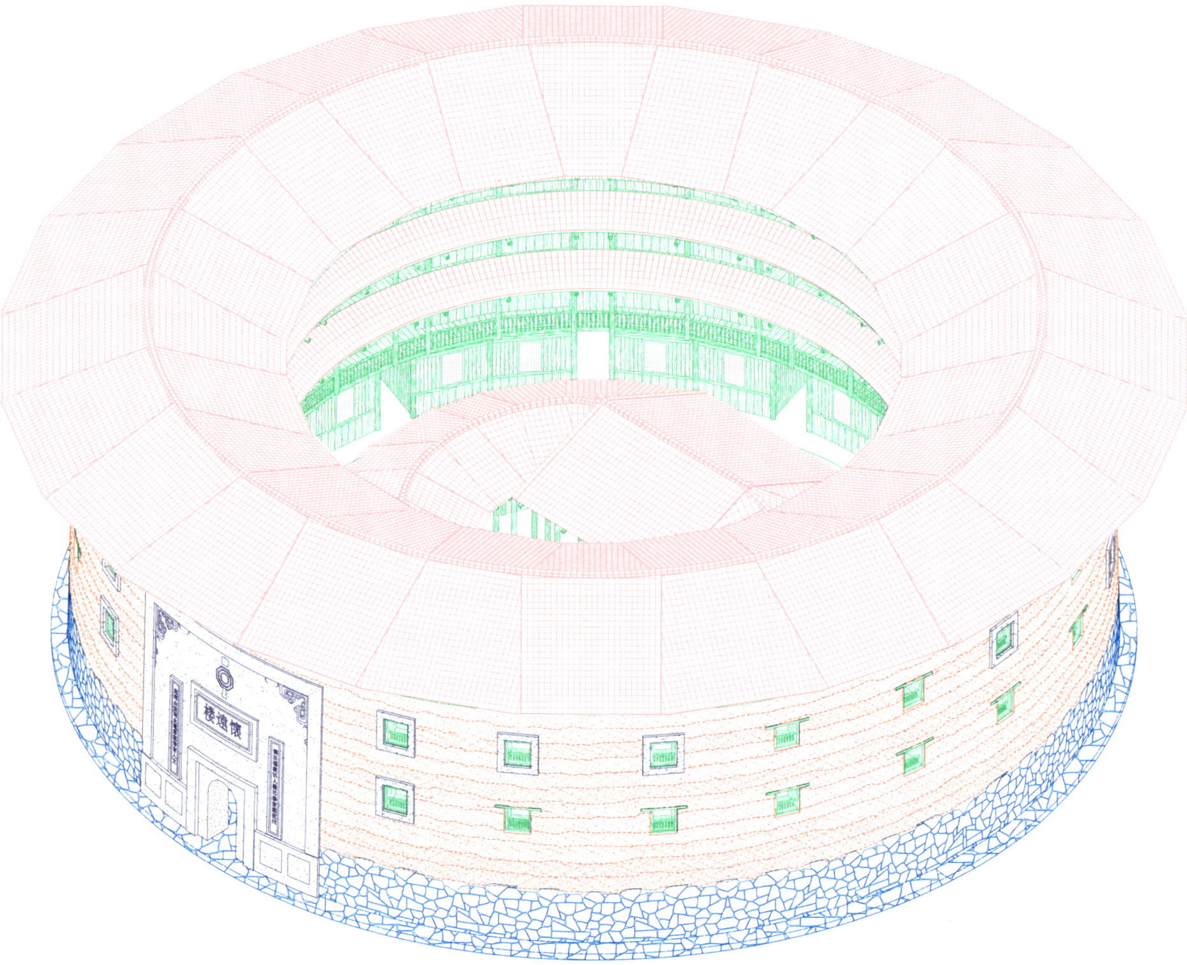

Earth, Wood, Stone,
Ceramic, Concrete

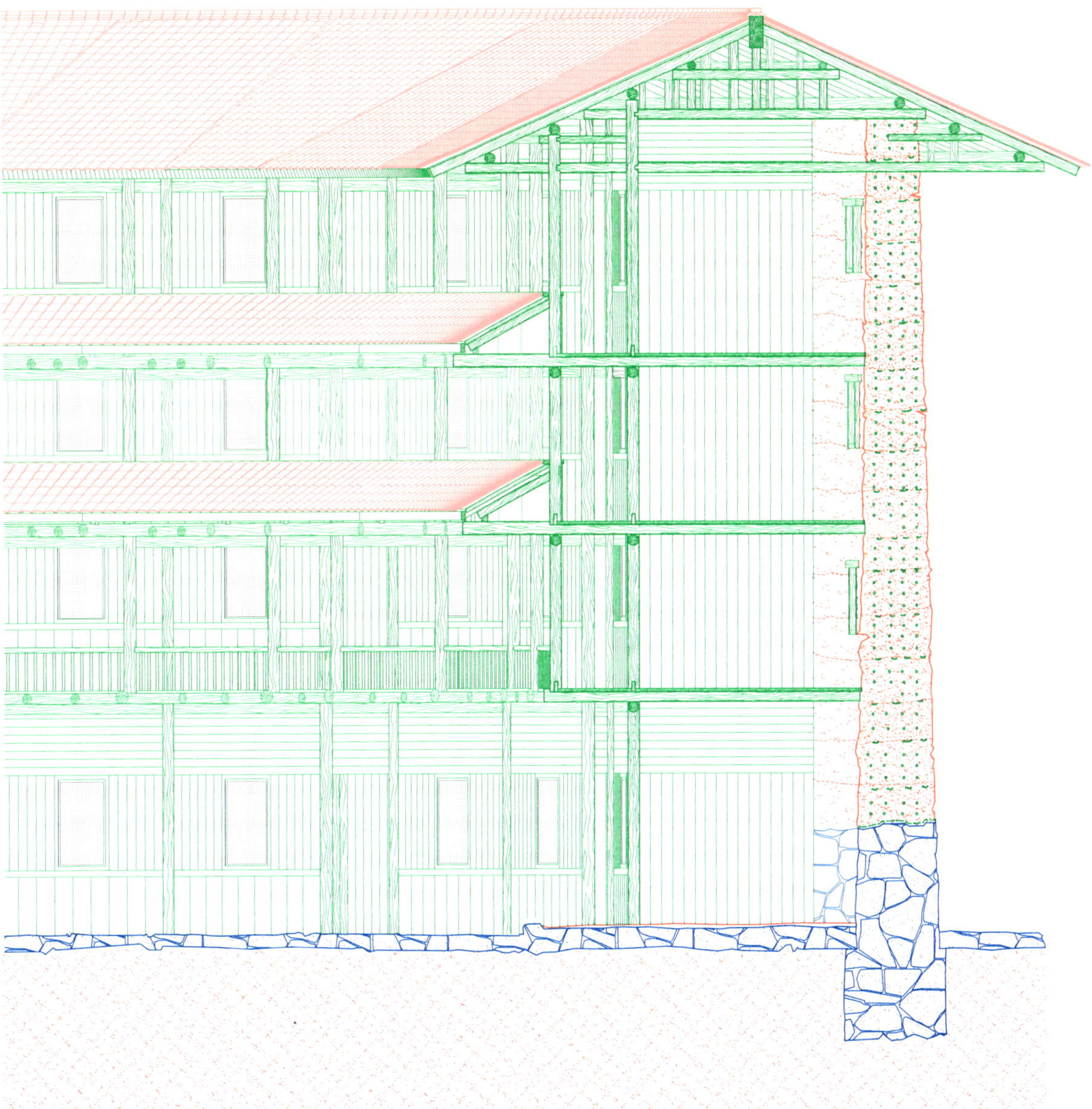

Name	UMGEBINDEHAUS
Location	Germany, Czechia, Poland
Year	15th to 19th centuries
Construction	Wood and Earth and Stone as Collage

This house type is found in a region that spans Germany, Czechia, and Poland. It takes its name from *umbinden* (German for "to tie around") as its ground floor is bound by a separate structure that supports the upper story and roof. The ground floor consists of self-supporting volumes that form a stable and living quarters: the living quarters are often of massive log construction, while the stable and foundation are of field stone masonry. The upper story and roof are a half timber framework structure infilled with plant-reinforced earth that sits on stone foundations and is independent from the main volumes on the ground floor. Reasons for this discrete mix of construction techniques are not well understood, but current explanations focus on differential movement between the stories. The graphic combination of various construction techniques—the horizontal stripes of the log cabin, the ornamental posts and diagonals of the timber framework, the field of clay roof tiles—create a rich overlay of structural, ornamental, and cultural meanings.

Wood, Earth, Stone,
Plants, Ceramic, Concrete

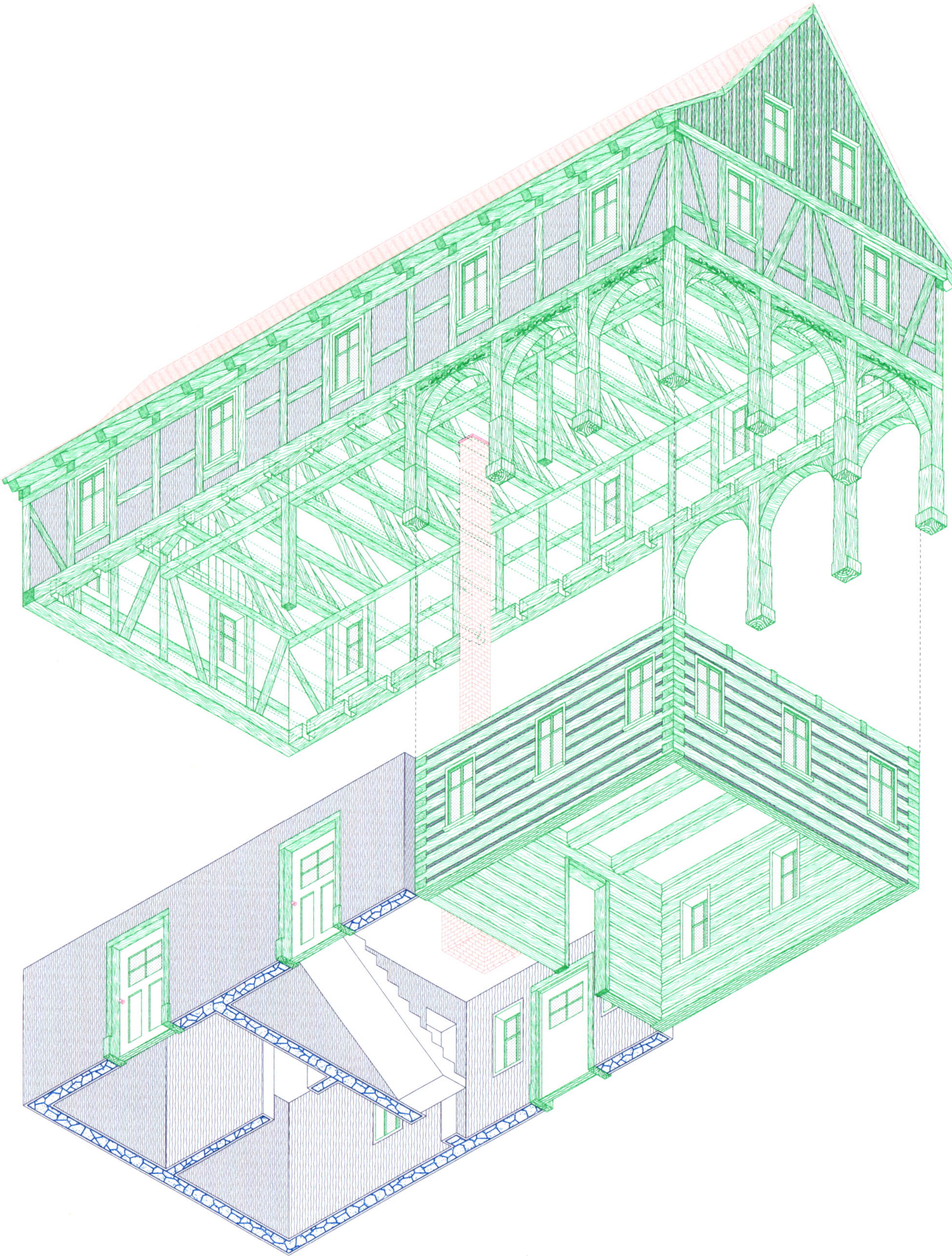

Name	YURT
Location	Central Asia
Year	Since 600 BCE
Construction	Textile and Wood as Wrap

The components of this construction type are easily deployable, transportable, and demountable in support of the inhabitants' nomadic lifestyles in Central Asia. The components of the Mongolian variety include ornamental central wood posts to support the center ring, a lightweight wood lattice wall at the lower roof perimeter, individual wood joists spanning from roof center ring to the perimeter, rope or leather ties for local connections, a tension ring around the top of the wall lattice at the base of the roof, and miscellaneous fabrics and hides are applied to the frame according to material availabilities and climate. The components are discretely layered in a manner similar to modern construction. However, unlike the contemporary case of construction with layers of discrete products driven by commercial efficiency, these layers are interdependent and integrated into a singular whole.

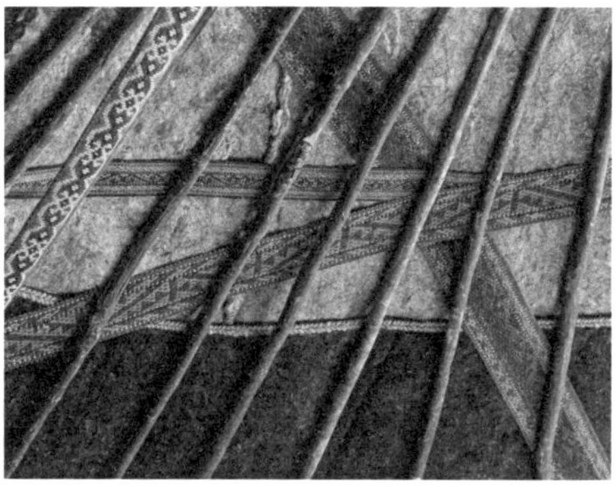

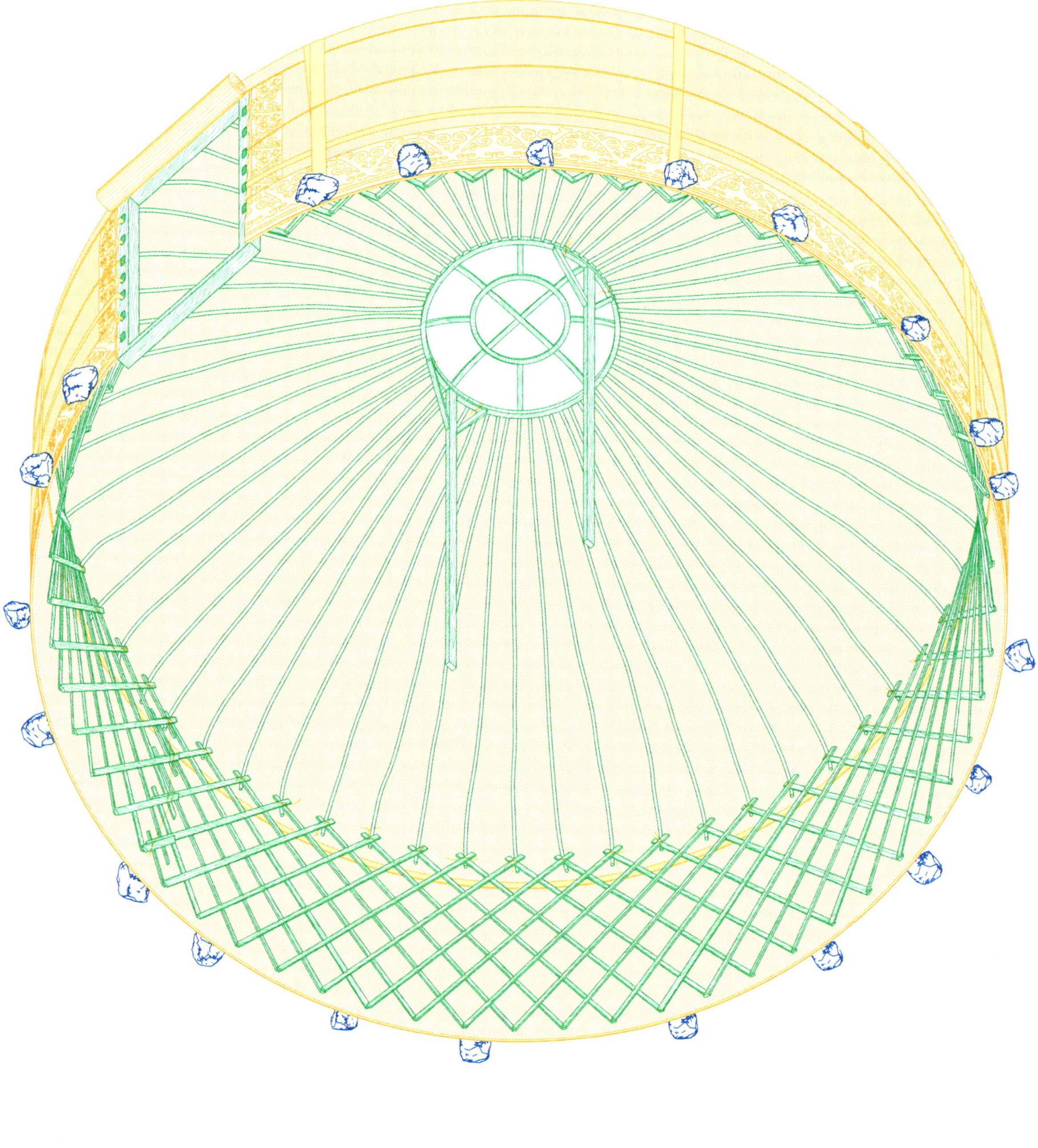

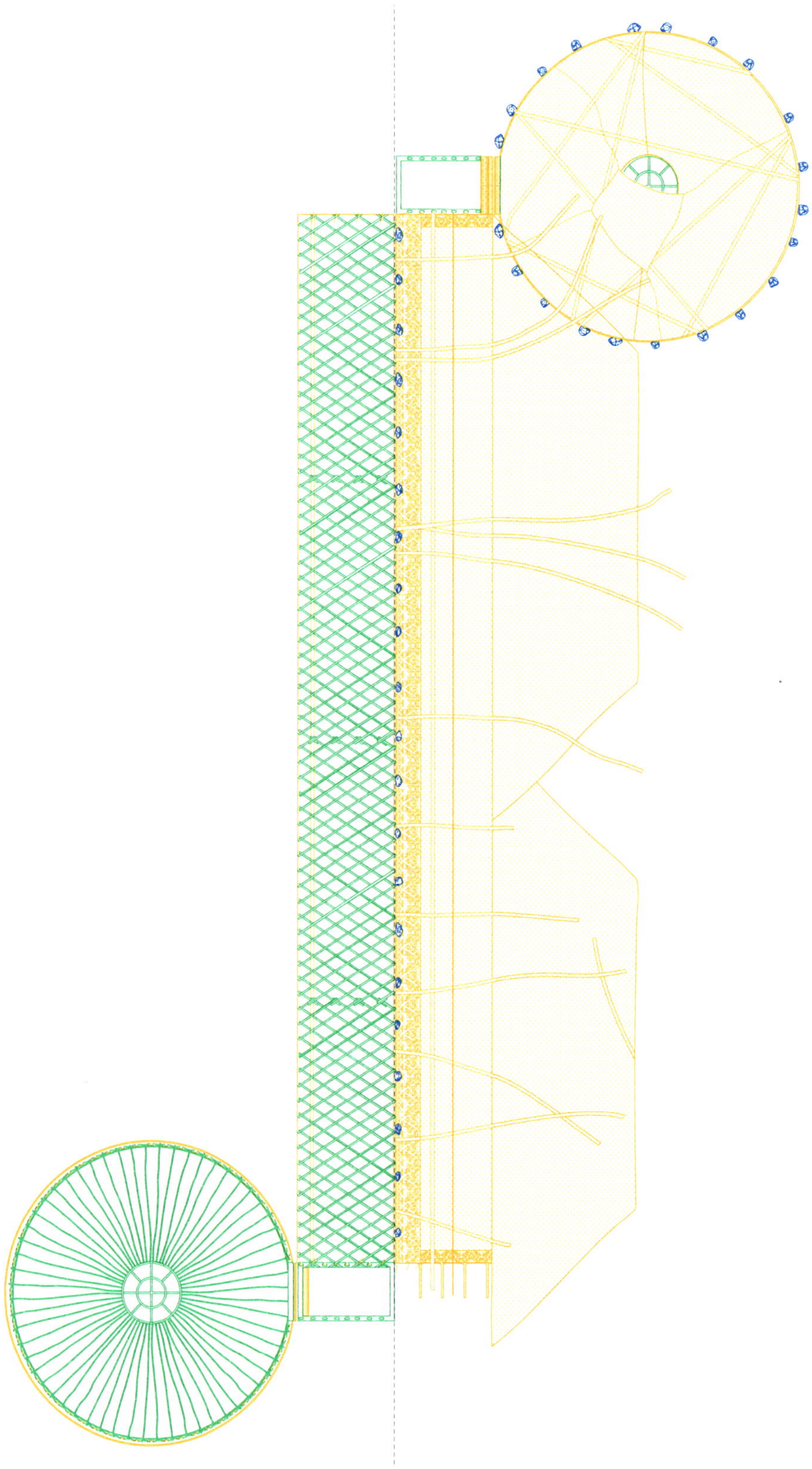

Aaron Forrest

Making sense of the heterogeneous through some kind of systematic analysis necessarily requires the creation of categories, which immediately present both conceptual and practical problems to the would-be categorizer. Conventional approaches to the categorization of buildings immediately fall short: organizing by structural system, construction method, material, or code-determined type are in turn found to be either inapplicable, impossibly narrow, or exclusive of the concept of heterogeneity altogether.

Upon a cursory examination of one of the many technical manuals of construction available to architects, one immediately finds building components broken down into functional layers that are then subdivided by material.[1] The designer interested in mixing these layers and materials is forced into a bind: they must either work with a language of purity well aligned with established structural materials and methods or employ a language of heterogeneity as a scenographic layer on top of an otherwise conventional structural skeleton.

Though the aforementioned manuals allow building designers to work rapidly according to convention with predictable results, they, through the same mechanism of expedience, foreshorten the horizon of possibilities available to architectural and structural designers. In an epoch increasingly defined by ecological catastrophe that finds its roots largely in these exact mechanisms, these limits—simultaneously mundane and epistemic in their scope—bar the designer from any alternative path. The regime of material classification that forms the basis of our field is now a barrier to future ecological and creative practice.

Other ways must be found, and similar questions confronted by parallel disciplines can be instructive. In his brief essay “The Analytical Language of John Wilkins,” Jorge Luis Borges discusses several proto-encyclopedias organized in apparently nonsensical ways.[2] The most famous of these, an apocryphal Chinese volume with the title *Celestial Emporium of Benevolent Knowledge*, categorizes animals according to the following schema:

(a) those that belong to the Emperor
(b) embalmed ones
(c) those that are trained

(d) suckling pigs
(e) mermaids
(f) fabulous ones
(g) stray dogs
(h) those that are included in this classification
(i) those that tremble as if they were mad
(j) innumerable ones
(k) those drawn with a very fine camel's hair brush
(l) others
(m) those that have just broken a flower vase
(n) those that resemble flies from a distance

This classification is so challenging to modern eyes precisely because it foregoes any single datum for comparison. Within a single table, categories range from entirely subjective to self-enclosed, from the singular to the impossibly vague. More importantly, there are no distinct boundaries between the categories. A particular animal (let alone an entire species) could easily fall into half the categories, and there is no animal that would not fall into at least two. Globally established categorical distinctions are nonexistent, in favor of locally—whimsically—determined identities.

The classification method Borges describes relies on empirically determined categories over any nod to idealism, to such a degree that it borders on nonsense. But if we take it seriously for a moment, its simultaneous elasticity and razor precision undeniably allow the reader to understand the qualities of an animal—as well as its relationship to other species—along several different axes at once. The higher level of complexity that the table allows is intriguing at the least.

Returning to architecture, we must acknowledge the chaos that Borges has identified here. An entirely empirical, locally determined classification system clearly falls short of establishing a true discourse among the objects under its umbrella. In the context of a collective activity like building construction, a system of material classification requires an internal consistency that allows materials and working methods to be discussed, shared, and proliferated among the myriad actors in a building's production. But the multidimensional qualities of the Borgesian scheme and its ability to see relationships between categories as complex and relational clearly hints at a model for understanding construction material that could be more open to the found, recycled, poor, and compound materials currently excluded from present classification models.

A more practical model might come from materials science, a discipline as impure as our own that deals directly with physical matter, though at a very different scale.[3] At the most basic level, materials science identifies only three major material categories: metals, ceramics, and polymers. Every material (whether elemental or compound) falls into either a subcategory or a blend of these types. An individual material's categorization depends largely on chemical composition and position on the periodic table, with each type generally (but not uniformly) exhibiting broadly similar characteristics in terms of ductility, conductivity, brittleness, etc. But even at this broadest of levels, some strange combinations appear. For example, wood and plant fibers are grouped alongside plastics due to their similar chemical structures.

Since materials scientists are typically more interested in the performance of materials than their appearance or correctness, they focus on seeking and inventing new compounds based on their suitability for a particular task rather than their adherence to a particular regime of categorization. Indeed, classification is a first-pass sorting mechanism rather than a final determination of a material's fate, and there is a general awareness that performance characteristics are often unstable (depending on environment) and that the boundaries between types are inherently fuzzy.

Because of these fuzzy distinctions, and with the range of available materials numbering in the tens of thousands, metallurgist Michael Ashby published the widely used Ashby plot in 1992 as a graphical means of assisting designers, manufacturers, and fellow materials scientists in narrowing the range of task-appropriate materials based on combined performance characteristics across multiple axes.[4, Fig 1] Ashby plots identify specific materials as points, and categories of materials as regions. When mapped together, they begin to make clear how different material types perform across the designated criteria individually, as a class, and relative to one another. The wealth of data they provide in a single snapshot vividly illustrates the shifting borders between metals, ceramics, plastics, and natural materials.

The precision with which Ashby plots map out the fuzziness of the world of materials has proven incredibly useful. The focus on utility has led materials scientists to invent materials with characteristics that architects could only dream of—ceramics that act like metals, metals that behave like glass, and materials that can jump back and forth between different guises depending on the application of light or electrical charge.

Of course we should not fool ourselves to think that materials in architecture are capable of the same fluidity as chemical compounds manipulated at the molecular level. Buildings, in the end, are made of much more brutal ingredients: industrially processed and mass-produced mud, rocks, plants, metal, and plastic. What we can learn from materials scientists isn't the appropriation of the latest supermaterials but the irreverence and conceptual flexibility with which they define and seek to understand the materials they work with.

Following this example, a true contemporary architecture of heterogeneity would value materials not for their individuality but for how they work together and how they can be alchemically transformed to act in very unexpected—and unexpectedly valuable—ways. Though there are few current architectural examples of this approach,[5] two Mies van der Rohe projects begin to point at the combined structural and aesthetic potential of this way of thinking. The first of these is the Barcelona Pavilion of 1929, whose hovering roof, slender columns, and free-floating interior partitions did as much to define modernist principles as any text. But the separation of structural and spatial elements in this example is rhetorical: the steel-framed roof appears to sit only on the columns, but in fact relied on temporary wood shoring for support during construction until the slender stone-clad partitions, which sandwiched steel props, unreinforced masonry, and drain pipes could be inserted beneath.[Fig 2] As Dietrich Neumann has noted, the reality of the structure was much more of a hybrid than legend would suggest.[6] But if we read the building as it stood—that is, as a simple plane innocently sitting upon a well-composed grouping of chrome columns and stone slabs—each material in the composition takes on a more complex and relational valence: each has a hybrid role to play structurally, symbolically, and in the definition of space.

The second example, the Golf Clubhouse in Krefeld (1930), existed only as a competition proposal until it was reconstructed as a temporary pavilion in 2013 by Belgian architects Robbrecht en Daem.[7, Figs 3, 4] The reconstruction capitalizes on the lack of definition in the original design to propose a hybrid steel-and-wood construction that acts simultaneously as a building, a model, and a mock-up. In lieu of steel, concrete, and exotic stone and glass partitions seen in Mies's contemporaneous built work, the reconstruction is built in a provisional manner, with a conventional (and easily repurposed) wide flange steel and wood superstructure supported on chrome-clad steel columns, with interior partitions made from stud-framed plywood paneling. Beyond the jarring shift in materiality, the construction is purposely left unfinished: areas left undetailed in the original competition entry are left without cladding; bare steel framing, wood studs, and freshly graded earth expose the messiness of design and construction processes generally absent from Mies's executed work.

Though somewhat accidental as models of heterogeneity, these two pavilions, built nearly a century apart, are suggestive of a conceptual fluidity around the functional role of materials in building construction that might serve as a basis for future thinking around architecture as a test bed for experimental construction. Each example can be read as a kind of material impostor attempting to translate a constructional idea between conceptually incompatible media: from column to wall; from steel to wood; and from luxe to poor. This act of translation, whether circumstantial or by design, forces a double reading, blurring the functional distinctions between materials that were so critical to the author's intent.

Translation, of course, is a technique that is deeply ingrained in contemporary architectural thinking. But in today's digital world, it is nearly universally restricted to the sphere of representation. And though its representational role is unambiguously valuable, a conceptual fluidity that remains safely in the realm of the image can only challenge how we perceive but not how we build, ultimately resulting in the reinforcement of existing construction standards. What the two Mies pavilions suggest is the idea of reinstituting construction—particularly experimental construction—as part of a feedback loop of design that runs from the sketchbook to the studio and to the construction site. This approach, which would require reconceiving construction as a process of experimentation, promises to interrupt our seamless modern design processes with some of the productive ambiguity seen in the literary and scientific examples and thus set the conditions for a new, heterogeneous present.

1 See, for example, Francis D. K. Ching, *Building Construction Illustrated*, 6th ed. (Hoboken, NJ: John Wiley & Sons, Inc., 2019); or Charles George Ramsey, Harold Reeve Sleeper, Robert T. Packard, and Dennis J. Hall, eds., *Architectural Graphic Standards* (New York: Wiley, 2016).

2 Jorge Luis Borges, *Other Inquisitions, 1937–1952*, trans. Ruth L. C. Simms (Austin: University of Texas Press, 1964), 101–5. For a deeper reading of the epistemic challenges posed by Borges' list, see Michel Foucault, *The Order of Things: An Archaeology of the Human Sciences* (New York: Vintage Books, 1970).

3 See Robert J. Naumann, *Introduction to the Physics and Chemistry of Materials* (Boca Raton, FL: CRC Press, 2009), 1–12; and Jennifer Mihalick, "Classification of Materials," from the unpublished manuscript "The Chemistry of Materials and their Role in a Sustainable Society."

4 Michael F. Ashby, *Materials Selection in Mechanical Design* (Oxford: Pergamon Press, 1992).

5 Though architecture remains mired in representational questions (see Yasmin Vobis's essay in the present volume), the parallel discipline of structural engineering abounds with innovation in this area. See, for example, the stress-ribbon bridges of Schlaich Bergermann Partner and the fabric-tensegrity vaults designed by Jun Sato.

6 Dietrich Neumann and David Caralt, *An Accidental Masterpiece: Mies van Der Rohe's Barcelona Pavilion* (Basel: Birkhäuser, 2020). The breakneck speed of design and construction led to a fascinatingly compromised final result. The reconstructed version of the pavilion, with its concrete slab roof and robust grid of steel supports embedded in the stone walls, is a much more conventional structure.

7 Christiane Lange, Robbrecht en Daem Architecten, *Mies 1:1 : Ludwig Mies van Der Rohe, the Golf Club Project / Das Golfclub Projekt* (Cologne: Buchhandlung Walther König, 2014).

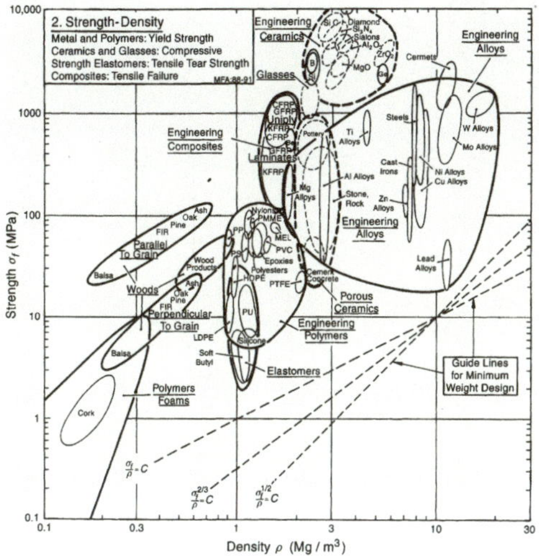

Fig 1

Fig 2

Fig 3

Fig 4

Fig 1 Ashby plot from *Materials Selection in Mechanical Design*, Michael Ashby, 1992.
Fig 2 Ludwig Mies van der Rohe, Barcelona Pavilion under construction, 1929.
Figs 3, 4 Robbrecht en Daem (after Mies van der Rohe), Golf Clubhouse in Krefeld, 2013.
Photos by Joachim Brohm.

The Difficult Imperfect Whole

Jeannette Kuo

> "The difficult whole in an architecture of complexity and contradiction includes multiplicity and diversity of elements in relationships that are inconsistent or among the weaker kinds perceptually."—Robert Venturi[1]

The drive for sustainability in the face of our climate crisis has brought architecture closer to constructive agendas as a means of defining a disciplinary language. While previous generations have foregrounded theoretical or geometrical pursuits as vehicles for architectural experimentation, the current era, finding itself confronted with decades of laissez-faire extraction and consumption, has turned to a more direct understanding of architecture as material construct with finite resources and specific social, economic, and climatic contexts. The intellectual and the physical have realigned to prioritize material performance and expression not only as means of addressing ecological responsibilities but also as grounds for experimentation, putting the spotlight on assemblies that we have long taken for granted. Construction isn't simply the translation of ideas into physical form but is once again considered the catalyst for a new design attitude where material compositions are exposed, interrogated, and celebrated. It is in this trajectory that heterogeneous construction has once again found a fertile platform.

Heterogeneity has historically been a means to be resourceful by combining performative qualities of different materials. It accepts that no single material is able to do it all and embraces construction as an assembly of parts. In this system of part-to-whole, construction is a constant negotiation between materials that resolve at joints and reveals. Whether as preemptive measures for differential movement and weathering or for the controlled aging of elements, the tectonic expression is a means to anticipate these unavoidable changes over time. Yet these imperfections or incompatibilities are often suppressed or even concealed behind an abstract surface. If we were to instead embrace construction as a reconciliation of disparate parts, it could open the door to a new language of ecological design in which the tectonic plays not only a spatial or haptic role but also a temporal one, extending the life span of a building. It would allow us to think about the hierarchies of material systems that allow for adaptation, maintenance, or even disassembly and recyclability.

In architectural theory the most prominent approaches to heterogeneity have dealt less with construction than with image and representation. Robert Venturi's *Complexity and Contradiction* offered up the breadth of architectural history as material for a combinatorial way of thinking and designing. It liberated architects from the weight of history while simultaneously acknowledging it. But much of that discourse is centered on the manipulation of image, representation, and symbolism. Since then, tectonic theories have been scant or even apologetic, acknowledging performance but shy in staking claims to an expressive language. Yet Venturi's text can be read in multiple ways: the representational is only one way of interpreting it and, I would argue, an oversimplification of his position. Today, it seems just as plausible to extract ideas of complexity and contradiction for a constructional way of thinking.

Heterogeneous construction is nothing new. Throughout most of human history, mixing construction was the everyday not the exception. In the era of pre-professional construction, heterogeneous construction was a form of pragmatism that allowed builders to adapt to exigencies. To take just one example, half-timbering was a means of building taller and denser with limited tools that allowed villagers to construct without specialized professional knowledge. The technique can be traced to sites in ancient Pompeii, where timber frames would receive brick infill. Structurally, the two materials act in tandem, one reinforcing the other, achieving what each alone cannot. But out of the method a certain language of heterogeneity emerged that is still present in Europe today. These lowbrow techniques could be argued as nonintellectual—as construction driven by the limitations of context. Alternatively, they can be read as creativity in the face of constraint; a means of acknowledging material imperfection that liberates designers and builders to work with a palette that is not pre-engineered to be flawless but instead is tied to local ecological and economic cycles.

The first instance of a style in Western history consciously built on architectural heterogeneity can be found in the unrealized works of Eugène Viollet-le-Duc, who left an immense trove of his thoughts in the detailed treatises of the *Entretiens*. These include the designs for the grand Concert Hall for Three Thousand (1864), in which iron struts support masonry balconies and vaults like spindly bones holding up an impossible weight.[Fig 1] Or in the Hall with a Twenty-Meter Opening (1865), in which delicate steel struts and iron tension rods take the place of volumetric squinches and pendentives that would have traditionally transferred forces from a domed ceiling to load-bearing walls.[2, Fig 2] Unlike his contemporaries, who experimented with fully iron-based systems that would then be clad in other materials, Viollet-le-Duc consciously played with the material contrast of masonry and iron, using the iron not as a complete replacement of masonry but rather in dialogue with it. The selective moments in which iron is used highlight its impossible strength and perceptual lightness. His attitude of sufficiency, using only as much as is needed and no more, seems apt for many of our contemporary discussions. But Viollet-le-Duc's call for structural expression was as much artistic as it was philosophical and technical. It is the integration of the three modes of thinking that sets his approach apart as a search for a new architectural language.[3]

A built contemporary of his work can be seen in the Menier Chocolate Factory,[p 80] a hybrid iron-and-masonry structure for a chocolate mill straddling the Marne River.[4] Designed by the little-known Jules Saulnier in 1869, the structure reinterpreted half-timbering in wrought iron. The lacelike iron structure infilled with brick was an exoskeleton that navigated not only material heterogeneity but also structural heterogeneity. The floors inside did not transfer onto solid ground but instead connected back toward the facades, where the loads were transferred across the surface onto girders that spanned between piers. Composed of discretely expressive structural parts, the building performed as a unified whole. The tandem performance of the iron with the infill renders the structure excitingly ambiguous and difficult to categorize as a structural type. We are left with an expressiveness that appears impossibly light and playfully decorative. Heterogeneity, like in Viollet-le-Duc's designs, was a means to manipulate material perception.

Today, heterogeneous construction is once again coming to the fore as a means of expression. As we confront the depletion of our material resources and the ravages of modern construction, hybrid approaches offer a means to redress the excesses of the past, enabling improved performance through strategic material consumption. Heterogeneity allows us to make the most of each material, celebrating their unique properties to produce a greater whole. They allow us to minimize the use of carbon-heavy materials like concrete and steel while improving the performance of natural materials like timber and clay by leveraging their unique characteristics. Further, heterogeneous construction offers a way to extend the lives of existing buildings whether through adaptive reuse or through renovation. They allow us to introduce new constructive elements into an existing system to produce a new whole. Like many lessons of frugality from past generations, heterogeneous construction is a means to be sustainable: to use what we need and no more.

Two architectural offices who take this position but from very different aesthetic approaches are Architecten Jan De Vylder Inge Vinck (AJDVIV) and Roger Boltshauser. One could argue that they are both results of operating within very particular contexts. In the case of AJDVIV, their conscious choice to design within existing buildings in the Belgian suburbs has led them to develop a style that is based on a strong methodology. The layering of new onto old happens at multiple scales: with the insertion of a colorful lintel to cut open an existing wall or the wrapping of a new thermal layer over that of a less performative one. These precise yet lyrical operations don't suppress the existing. Instead, they acknowledge the tandem role of new and old. The projects seem to suggest that the built is but a frozen moment in time, open to future adjustments and modifications and therefore without end; that coexistence, including the imperfections in between, creates a new—if incomplete—whole.

The work of Roger Boltshauser, a Swiss architect who has emerged in recent years as a prophet of sustainable construction, has focused on building techniques that—combined with modern-day technology—can revive traditional bio-based materials. His experiments with rammed earth walls,[p 74] first in his own atelier and later in much larger projects, combine the very rough and structurally weak loam with concrete and steel to achieve greater height and stability. Unlike the intentional collage of his Belgian counterparts, Boltshauser's work attempts to resolve toward a recognizable whole but always with a certain tension between its composite parts; the archaic expression of the rammed earth is pitted against industrial elements whose sharpness hints at their engineered precision.

The demand for precision in modern construction in combination with the unpredictability of many bio-based materials such as rammed earth or timber is front and center in many constructive experiments today. This was a problem recently faced in my office, KARAMUK KUO, with one of our projects, a speculative lab building in Basel, Switzerland.[Figs 3, 4] While we were interested in using a sustainable timber structure, the sensitivity of the chemistry instruments to be used in the building ruled out any system prone to vibrations, effectively constraining us to in situ concrete. Faced with this contradiction, we developed a hybrid system together with Schnetzer Puskas Engineers that would satisfy both the stiffness required while using the building as a carbon sink. In a concrete building, the largest volume of material used isn't in the vertical structure but in the horizontal slabs. The greatest opportunity for reducing concrete use is therefore in rethinking the floor plates. In our case, to maintain the stiffness of a diaphragm we developed an in situ concrete frame—essentially a horizontal Vierendeel—where the voided fields would be plugged with mass timber infills. This hybrid system not only allowed us to reduce the use of concrete by roughly 40 percent per slab but also addressed two other considerations: the long-term flexibility of the building and its spatial expression. The use of concrete is amortized through its durability, further complemented by the removable timber infills that allow future tenants to connect multiple floors without risking structural integrity, ensuring the adaptability of the building for the unforeseen. Going further, the hybrid system develops an expressive potential very often ignored in a slab. The lab environment, typically sterile and artificial, gains a visual comfort and domesticity with the exposed timber and concrete structure, the two materials striking a balanced rhythm between structural pragmatism and expression.

As contemporary practitioners wrestle with design approaches inherited from the past half century, one thing has become increasingly clear: the greatest myth of the modern period is the myth of perfection. With the standardization of architectural products, we have grown complacent in how we consider materials and assemblies and in the ways that we interact with our buildings. We expect that materials will be preprocessed to achieve standards that are often unnatural. We expect them to have durability that defies biological time.

We expect spaces to last forever with minimal maintenance. We expect joints and seams to disappear so that we can have pristine abstract surfaces. The list goes on.

And yet the material world that we live in is innately imperfect. No material can do it all. All materials age. And if they don't, we should probably question their provenance and manufacture. Starting from the assumption of imperfection allows us to embrace the complexities and contradictions that come with construction. While approaches can vary dramatically from Viollet-le-Duc to AJDVIV or Roger Boltshauser and our own work at KARAMUK KUO, this lineage of design thinking accepts two common premises: that the performative and the expressive do not have to be different; and that the challenge and beauty of our discipline lies in the negotiation of imperfection.

1 Robert Venturi, *Complexity and Contradiction in Architecture* (New York: The Museum of Modern Art, 1977), 88.

2 *Viollet-le-Duc* (Paris: Editions de la Réunion des musées nationaux, 1980). This publication accompanied an exhibition at the Grand Palais in celebration of the centennial of Viollet-le-Duc's death and is one of the most extensive publications on his oeuvre. It presents unbuilt works of his within the context of his peers, as well as the impact he has had on subsequent generations.

3 *Viollet-le-Duc*, 252–53.

4 *Viollet-le-Duc*, 370. The Menier Chocolate Mill was featured in the exhibition catalogue on Viollet-le-Duc as one of his few contemporaries who instrumentalized engineering and performance as a means of expression.

Fig 1

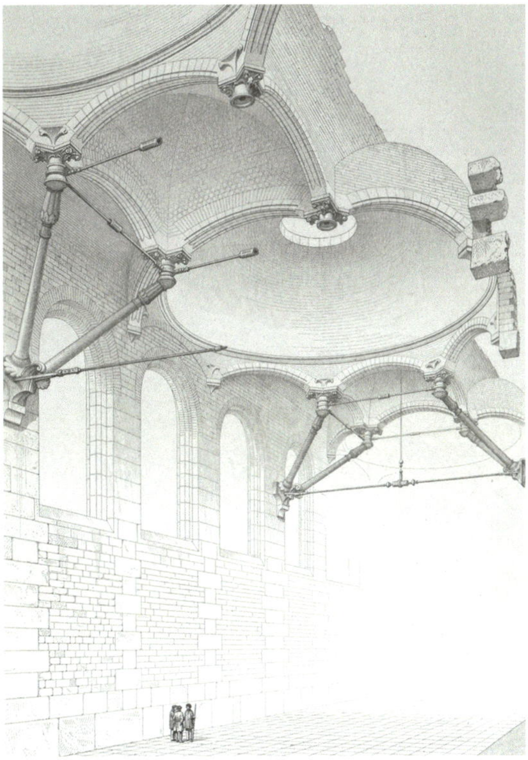

Fig 2

Fig 3

Fig 4

Fig 1 Concert Hall for Three Thousand, Eugène Emmanuel Viollet-le-Duc, 1864.
Fig 2 Hall with a Twenty-Meter Opening, Eugène Emmanuel Viollet-le-Duc, 1865.
Fig 3 Research Lab Building, collective ground floor and mezzanine, KARAMUK KUO, 2021–29.
Fig 4 Research Lab Building, atrium with view to labs, KARAMUK KUO, 2021–29.

Yasmin Vobis

Mixed media became a fundamental art medium in the twentieth century, forming the basis for much of avant-garde art production to this day. It is curious to consider then, that modern architecture has shown a stubborn resistance to engaging its mixed material roots. The "all-steel," "all-concrete," and more recently "all-wood" building have each become entrenched structural and aesthetic paradigms that have dominated architectural practice but whose purist preconceptions proscribe a more complex interplay of image and construction found in mixed-media art practice. The Examples and Prototypes presented in this volume ask the question: How can architecture become self-conscious of its heterogeneity again? In search of an answer, it may be useful to revisit a counterlineage to modernism's purisms, one more closely aligned with concepts in art practice that may offer an alternative path forward.

Performance and expression drifted apart in modern architecture in the mid-twentieth century, cementing a set of attitudes that still shape construction today. It has been well established that the increasingly complicated technical demands of the building envelope were solved by separate control layers to address each function: a layer for insulation, waterproofing, moisture regulation, exterior protection, cavity ventilation, interior finish, and so on.[1] This progression fundamentally separated building performance from the aesthetic roles of construction elements over time. The typically homogenous finish of the wall was reduced to the outermost of these layers, transformed into an unencumbered representational role while concealing all of the aesthetic, social, and ecological consequences of the wall's internal layers.

Monofunctional layering constituted a conventional framework that became entrenched dogma in the decades following World War II. These conventions were reinforced from all sides—by codes and industry standards in practice as well as in schools through academic curricula that emphasized standard construction systems that could be added together ad infinitum. Postmodern architecture took little risk in relation to these base assumptions: although it played with the aesthetic image of heterogeneity (think of the formal heterogeneity championed in the work of Venturi Scott Brown as one example), the resulting buildings blindly accepted the conventional layering of materials and furnished abstract finish surfaces that purposely masked both the appearance and technical role of the wall's internal construction. Today, the layered paradigm is further exacerbated every time new performance criteria arise through the addition of ever more layers. As a result, architects have ended up in the bizarre situation of having to hold together and be accountable for this increased complexity as well as having to invent an additional image—a representation—out of thin air, to be layered over the top of the materials of construction.

Frank Gehry's projects in the 1970s and early 1980s represent an important but isolated break from this line of development. Rejecting modernism's aesthetic of purity, Gehry's experimental combinations of lowbrow materials developed an aesthetics based in material difference. Surfaces in projects like the Gehry Residence (1977)[Fig 1] or Indiana Avenue Apartments (1981) read alternately as assemblages of disparate materials or as the layers of conventional construction delaminated and put on display. Structural studs are laid bare; plywood becomes visible and takes on a new role as exterior cladding; asphalt roof shingles are displaced to vertical facades; and corrugated metal and chain-link are transposed into the domestic sphere. This self-described "cheapskate architecture"[2] transformed poor, cheap, and mass-produced materials into signifiers of cultural value, pushing materials' functional role to the surface while simultaneously making explicit references to other contexts in which they are more familiarly seen: the warehouse, the lumberyard, and the schoolyard.

Gehry's architecture closely paralleled the American art scene during this time, and it is in the work of his artist friends that Gehry recognized the freedom to work in such a multivalent manner. Unlike architecture, certain strains of art practice during this time pushed for a more complex understanding of material relationships to produce meaning. Rejecting the strictures of medium specificity, a number of midcentury American artists drew on prewar experiments with collage and assemblage to forge complex entanglements between media within single works. To take one example, in Robert Rauschenberg's *Combines* and later works, meaning develops in a nonlinear and relational manner through the combination of painting, collage, photographic images, and found objects. In works such as *Untitled* (1955),[Fig 2] there is no dominant medium; the found and synthetic elements work in tandem, adeptly pulling together references to diverse topics such as flight (a flattened toy parachute), the body (a found sock), a pastoral landscape (a postcard of cows in a field), and spatial depth (brushstrokes that at times suggest atmosphere while at other moments become self-conscious of their flatness). In comparison to pure painting, the expedience with which multiple references can coexist within a single mixed-media work is profound. The process of juxtaposition allows literal pieces of garbage to attain new meaning by being placed in a nonhierarchical relationship with other elements, painted and ready-made.

In a more contemporary example, we see architecture freely reassembled through the lens of mixed-media art practice in Lais Myrrha's installation *Dois pesos, duas medidas/ Double Standard* (2016).[Fig 3] The work presents the viewer with two constructions: each a pile of different materials as they are found in industrial and Indigenous construction. We are invited to compare two wildly different cultural paradigms through the language of concrete, bricks, steel, and glass on the one hand, and logs, straw, and mud on the other. The simplified format of the rectangular columns allows for a focus on the different materials rather than the overall form. In each, the form and space of a building is abstracted and made mute until all meaning and embedded knowledge reside in the differences between the construction materials, which embody the habits, cultural practices, and environmental attitudes of each culture directly rather than by way of representation or abstraction.

It may be worth pausing to reflect on how vastly different this approach to material is than in architectural postmodernism. In Leo Steinberg's foundational essay on postmodernism in art, he characterizes Rauschenberg's works as horizontal flatbeds rather than vertical, illusionistic picture planes: "To cope with his symbolic program, the available types of pictorial surface seemed inadequate; they were too exclusive and too homogeneous."[3] He goes on to liken the nonhierarchical ordering of elements on the surface to disordered desks and unswept floors, and finishes with "what he invented above all was, I think, a pictorial surface that let the world in again. ... As a criterion of classification it cuts across the terms 'abstract' and 'representational.'"[4] Steinberg's definition of postmodern art relies on the radical juxtaposition of abstraction and reference that develops out of an immersion in the multiple media of the information age, their construction, and the complex relationships of meaning that unfold between them. Architectural postmodernism, on the other hand, tried to create a complexity of appearance without any interrogation of its underlying media.

It is unclear why mixed-media art practice—and Steinberg's more complex dialogue between technique and appearance—has not impacted today's architecture more, given how the field has borrowed freely from developments in art in the past (consider the long and unchallenged reach of minimalism in our field). While Gehry clearly grasped its significance,[5] we are left to ask what form a contemporary approach to construction *as mixed media* might take. Certain strains of contemporary architectural production provide some hints, building on the promise of Gehry's work while retaining vestiges of the postmodern split between expression and construction. Both trajectories have gained new urgency in response to current global crises—with the hunch that the exposure of the material makeup of architecture can both lay bare and diagnose its wide-ranging ecological and social impacts.

The first of the two is a reworking of Gehry's aesthetic heterogeneity, where an emergent bricolage aesthetic tied to adaptive reuse acts as a response to architecture's throwaway culture as well as its long-standing aesthetic asceticism. We can see this approach in practices like Architecten De Vylder Vinck Taillieu (ADVVT). The aesthetic strategies from their early renovations permeate even ground-up buildings such as their Les Ballets C de la B and LOD (2008), which has the piecemeal aesthetic of a renovation. Variously sized lintels are highlighted as artifacts rather than seamlessly integrated. Interior walls appear unfinished. And the elements of wall construction are pulled apart on the main facade as though they had been layered over time by distinct authors: glazing units, shades, steel structural elements, and brick infill are made visible, their differences exaggerated by extreme coloration. Even though each material remains functionally separate, the extroverted visual heterogeneity of the project didactically exposes and improvises on the conventions of layered construction. While ADVVT's process is highly mediated by representation,[6] their aesthetics of difference make important inroads to embracing architecture's innate heterogeneity. Their musing "when things that have nothing to do with each other all of a sudden have something to do with each other"[7] is a call for architecture to be more radically open-minded about how different elements and materials may be brought into relation.

Working in parallel to these contemporary assemblagists, another set of contemporary practices and building scientists are reworking architecture from the inside out in search of performative heterogeneity. Technologists in this group have been focused on developing ways in which the materials in architecture can defy the strict functionalism of layered construction.[8] The research group Einfach Bauen (Building Simply) at TU München led by Florian Nagler notes that contemporary construction has become too complicated.[9] As a counterpoint, their experiments use oversized monolithic materials to absorb all roles of the building envelope in a single layer to reduce complexity: by reducing resource consumption and by simplifying construction and use, future reuse, and an accounting of its environmental impact. The project deliberately hybridizes and blends the performances of materials, asking each to take on roles that go far beyond their typical identities. For example, the thick monolithic walls of their lightweight concrete pilot project at Bad Aibling[Fig 4] blend structure, insulation, fire protection, acoustical separation, moisture barrier, and finish in a single material. The lightweight concrete walls are unreinforced, further simplifying the construction process and removing carbon-intensive steel; as a result, the punched windows take on arched forms to span over the openings, linking tectonic expression with structural behavior directly. In many ways, this approach rediscovers some of the earlier lessons of vernacular traditions, including a massive materiality that blends material performance and an entanglement of expression and construction, yet it does so through highly technological, contemporary means. But such insights are only possible through an experimental attitude intent on challenging prevailing construction paradigms.

Neither the approach of the assemblagists nor the technologists in itself provides a complete answer to the questions raised by heterogeneous construction. But taken together, the referential approach of the collagists in combination with the empirical rigor of the technologists offers a compelling path forward. In an era when architects repeatedly rush headlong toward new, singular supermaterials—these days bio-based instead of steel and concrete—it is perhaps time that we look more skeptically upon the field's tendency toward monomaterial solutions and instead look for intelligent and resourceful ways of designing the relationships between materials. New frameworks are needed, and mixed media provides a nimble analogue for an architecture with the sophistication to pull together questions of technical performance and cultural value. Following Steinberg's formulation of the flatbed picture plane, contemporary heterogeneous construction could productively resist the purist preoccupations of the modernist movement to relish in the constant material interplay between construction and image.

1 As an example, see Edward Ford's *Details of Modern Architecture*, which includes a set of three Louis Kahn wall sections that show the problem of pairing monolithic walls with the logic of modern layering. In addition, for a more detailed discussion of the development of layered construction, see Alejandro Zaera-Polo and Jeffrey Anderson, *The Ecologies of the Building Envelope: A Material History and Theory of Architectural Surfaces* (New York: Actar D, Inc., 2021). Contemporary construction manuals, such as Ed Allen's *Fundamentals of Building Construction* and Andrea Deplazes's *Constructing Architecture* teach the principle of layered construction as well.

2 This is Frank Gehry's own term taken from his lecture at Rice University, entitled "Cheapskate Architecture," in November 1976.

3 Leo Steinberg, "Other Criteria," in *Other Criteria: Confrontations with Twentieth-Century Art* (New York: Oxford University Press, 1972), 88.

4 Steinberg, "Other Criteria," 90.

5 It is no wonder that Gehry's work has resisted fitting in neatly with postmodern narratives of his time. His attitude to construction has always been more direct than the general tendency of American postmodernism toward representation, as seen in the work of Peter Eisenman, Venturi Scott-Brown, Michael Graves, and so on.

6 Representational techniques serve as conceptual models in ADVVT's work in the form of hand sketches, layered collages, and a long-standing collaboration with digital collage artist and photographer Filip Dujardin.

7 Architecten De Vylder Vinck Taillieu, *Architecten De Vylder Vinck Taillieu: Variete/Architecture/Desire* (Tokyo: Toto Publishing, 2019), 10.

8 While the terms *assemblagists* and *technologists* are an oversimplification of the depth of these designers' work, the categories help situate a growing set of practices that, as a constellation, are forging an urgent and critical dialogue between issues of expression and performance. We may look to the work of Boltshauser Architekten and BLAF Architecten as examples.

9 Frank Kaltenbach, "Building Affordably? Building Simply! Florian Nagler on Building Simply," *Detail Inspiration* (June 2022), accessed January 23, 2024, https://inspiration.detail.de/Download/document-download/id/6290d7caae98a.

Fig 1

Fig 2

Fig 3

Fig 4

Fig 1 Gehry Residence, Gehry Partners, 1978. Photo by John Nicolais.
Fig 2 *Untitled*, Robert Rauschenberg, 1955.
Fig 3 *Dois pesos, duas medidas / Double Standard*, Lais Myrrha, 2016.
Fig 4 Bad Aibling Prototype, TU Munich Building Simply Group, 2018.

Grid, Machine, Mud, Work, Air

Ajay Manthripragada

In the mid-nineteenth century, missionaries from Basel, Switzerland, set up a terra-cotta roof tile manufacturing industry along the Malabar Coast of India. Drawn to the dual opportunity of profit and proselytization, the Swiss entered the already manifold sociopolitical milieu of the region, in which Indian, Portuguese, and British interests intermingled. Manufacturing was but one of the mission's society-shaping efforts, which included the operation of schools, churches, and hospitals. The terra-cotta industry, in particular, experienced enormous success, with up to 60,000 Malabar tiles produced daily; they cover buildings throughout the region and far beyond. But it is the factories built along the coast, mainly around the port city of Mangalore, that most clearly express the disparate and even opposing forces at play in this regional economy. Seven were initiated by the Basel Mission and dozens more were founded by Indian merchants over the ensuing decades.

The factories stage several forms of material and immaterial heterogeneity. Each is a node in a vast and multifaceted territorial network of extraction, production, and shipping. Of the handful still in operation, Albuquerque & Sons is the oldest and largest. The company has remained continuously in the hands of the Albuquerque family, now in its fifth generation of ownership, since its Indian founder left the Basel Mission enterprise to venture on his own in 1868. Located in Mangalore near the water's edge, the main production facility is a diverse lattice in terms of both abstract organization and material composition, mapping the routines and rituals of work. Promising the total cooperation of space, manual labor, and inventory, the factory regulates the movement of employees alongside the flow of materials. Workers, roof tiles, and air weave through the building's structural grid, linking otherwise separate environments.

While decisions behind the factory's design were primarily pragmatic in motivation, the building exhibits an intensely unique appearance and atmosphere. Several material systems are intertwined in such a way as to reveal their individuality while obscuring their comprehension as a total system. Diagrammatically, the construction is derived from a generic European factory model. But built using local conventions and molded to the particular demands of tile production, a new type is produced altogether.

Two main sections comprise the factory complex. An original 1868 building, which housed three now-defunct kilns and their towering chimneys, sits adjacent to one built in 1986, with no spatial division between them. Both buildings share a gently shifting structural grid of hundreds of columns which seems to sprawl within an irregular bounding perimeter. Whereas the original factory was built with timber frame on masonry columns, the new, active portion is built with both masonry and concrete columns and a hybrid wood-and-steel assembly for the floors and roof. The entire complex is covered with pitched roofs clad with red tiles produced on-site, save for a light metal truss and polycarbonate covering that bridges the two sections. The plan of the complex[Fig 1] is reminiscent of the Great Mosque of Córdoba: an orientation-less array of columns interrupted occasionally by larger objects such as the kilns. Because the distinction between old and new is not readily apparent in plan, the geometric order of the grid can be read as both complete and endlessly expanding.

When one is immersed in the factory's interior, the overarching logic disappears. Contrast and difference between things are more apparent than their calibrated integration. Representative elements of the construction are visible, but when occupying the same space as complex machinery and equipment—generators, presses, mills, kilns, conveyors—their role beyond the immediate space is difficult to discern. As the required volume of space varies in length, width, and height from one operation to the next, the building is also spatially diverse, masked by the uniformity of the exterior and amplified by the unique cacophony of each zone. Overlaid onto this controlled chaos is more sensorial complexity. Breezes carry smells from the raw clay, oil used to lubricate equipment, and nearby fisheries. Sunlight shines through from multiple sources, intended and not. The drone and clicking of machines is constant. In all, the effect is staggering in its lack of unity.

Isolating the principal area of the three-story new building offers some clarity. A 36-chamber masonry kiln, arranged in two parallel rows of 18 chambers each, is the central spine of the ground floor, supporting and determining the structure above. Each chamber measures 3.7 × 4.5 m in plan, which directly corresponds to the spacing of the column grid of the building. Adjacent to the kiln is a line of press-molds, where tiles are formed from processed blocks of clay. The second floor, used for various functions including the storage of cashew shells for fueling the kiln, is made up mostly of the brick-paved top of the kiln itself, with a perimeter structure of timber joists. Vents in the top of the kiln are opened on the second floor during firing and the subsequent cooling, allowing hot air to travel up to reach the third floor, where the next batch of freshly molded clay tiles is drying. All three floors, and indeed most spaces in the factory, are connected by a network of conveyors endlessly moving tiles in all directions. Once bone-dry, clay tiles on the third floor move back down to the kiln for firing into terra-cotta, completing the loop.

The space of the drying racks—ostensibly one giant room—is the most sublime in the entire complex. An intricate system of steel shelving (the old factory used heavier wood shelves), stacked floor-to-ceiling and column-to-column, is meant to maximize air exposure to the drying tiles. Dimly lit and humid, the space is packed with a three-dimensional matrix of raw clay. This total homogeneity—an isolation of function and structure into a cloud of freshly formed tiles—is paradoxically possible because of the factory's hybrid structure.

The original 1868 building, while not in use for tile production since the 1986 addition, stood intact until governmental dredging of a creek along the edge of the property caused partial subsidence in 2012. An original pottery workshop remains in use, and some areas have been repurposed for storage, but most of the building is in a state of ruin. The 1986 building, on the other hand, is perpetually updated to support its ongoing activity. Notwithstanding the added effect this has on the apparent dissonance of the interior, the constant change reveals another salient aspect of the building's heterogeneity. Parts that have deteriorated over time have been replaced. The building is marked with such interventions throughout, sometimes crudely announcing themselves as substitutions. The geometric precision of the overall structure is offset by the textural roughness of these accretions.

Taken together, the registers of heterogeneity the factory exhibits—through use, space, material, labor, construction, geometry, time—offer a countermodel to a totalizing architecture conceived in full from the start. Instead of the stable image of the monolith, the Albuquerque factory demonstrates another state of equilibrium. Here, the demands that shape a building in its inception and the contingencies that unfold over its life produce a sustaining ethos rather than detract from a preconceived aesthetic concept. Such an outcome is only possible when control is tempered by and sometimes gives way to circumstance.

This acquiescence finds an analogue in the roof tiles themselves, produced in almost exactly the same way since the factory's founding. Balancing mass manufacture with handcraft, they display qualities of repetition from mechanized processes as well as idiosyncrasies from their handling, even to the level of individual workers' fingerprints. They are all identical and each totally different.

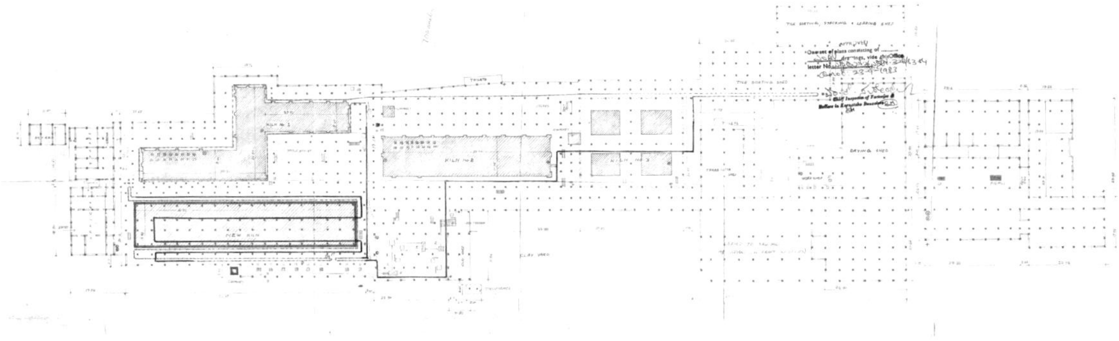

Fig 1

Historical information for this essay was obtained through conversations with the Albuquerque family and with the factory's manager, Raymond Monteiro, a fifty-one-year veteran of the company. For further insight into the Mangalore tile industry and its architecture, see a compelling series of essays written by Arijit Chaterjee and Asha Sumra, published online through *Architectural Digest India* and *Architexturez South Asia*.

The photographs accompanying this essay were made using a unique panoramic technique. With a shift lens and circular mount that enables the lens to remain in a single position while the camera sensor moves to capture peripheral information, a set of thirteen images was taken for each photograph and automatically stitched together. The stitching was not subsequently altered, revealing the limits of the original field of view. This technique aims to expand the reach of the photograph as much as possible. At the same time, it minimizes the distortion normally associated with wide-angle optics and eliminates the parallax problem of composite photography.

Fig 2

Fig 1 Plan of the factory complex with the newer portion at lower left.
Fig 2 View toward the 1868 portion of the factory including an old kiln, with new tile storage at right.

Fig 3

Fig 4

Fig 3 Triple-height space connecting clay processing section at left and kiln and drying floors to the right.
Fig 4 Third-floor drying room showing tile conveyor to the right.

How to Build Heterogeneously

Brett Schneider

How can we build heterogeneously? From the perspective of an engineer, this nominally translates to: how can we evaluate, quantify, and qualify the performance of heterogeneous constructions so that they can be viable in our contemporary construction and regulatory environment?

One answer is through analysis and testing leading to adoption in contemporary building codes.[1] For example, this is the method by which straw bale construction was adopted into the International Residential Code (IRC) as Appendix AS in 2013 allowing for its use in construction.[2] However, a closer look at this specific process reveals the limits and impracticality of this approach. Adoption required more than thirty years of development and negotiation including the reduction of this general type to a strictly standardized system with prescriptive requirements of use at limited scale. At the same time, building codes like the IRC also include provisions for "alternate materials, design and methods of construction and equipment"[3] that might suggest a more direct and flexible process to apply. In practice, these provisions tend to stifle rather than enable alternative applications, as approval depends on the opinion of a building official rather than any objective process, and it is unusual for building officials to have the specific expertise required to evaluate alternative design proposals. This all implies (at least in the United States as considered here) that the conventions of contemporary building codes do not provide an efficient path to expand building heterogeneously.

Before positing another path to address the challenge of this essay, it is worth considering the deeper reasons for the apparent incompatibility of heterogeneous construction and contemporary building culture. Let us step back to the introduction to this book, in which we consider the Menier Chocolate Factory[p 80] relative to the modern canon posited by Sigfried Giedion. As we note there, Giedion reduces the composition of this building to its iron frame to place it as a precursor to later discrete frame construction. Subsequently in *Space, Time and Architecture*, he follows a similar description with a lament that the building did not influence the development of skyscrapers in late nineteenth-century America as a discrete frame.[4] This commentary is important, as more recent research by Randolph Langenbach directly ties traditional infilled frame construction to the development of the Chicago Frame.[5] In one case, he argues, through reference to historical architectural journals, that William Le Baron Jenny may have been influenced by European examples of masonry-infilled timber frames for the use of masonry infill of iron frames in early skyscrapers.[6] Through this line of connection, the early skyscrapers were indeed similar to and likely derived from the heterogeneous combination of frame and infill we see in the Menier Chocolate Factory[Fig 1] and other vernacular examples considered in this study. The difference in these two narratives derives from Langenbach (unlike Giedion) looking at how these structures worked rather than constructing what can be termed a more art historical and formal narrative of architecture. Specifically, Langenbach references historical engineering publications[7] and accounts of the performance of buildings with infill frames in the San Francisco earthquake of 1906[8] to make clear that the use of masonry infill for resistance to lateral loads in early tall frame structures was both common engineering practice and understood to be highly effective in resisting seismic forces. In this frame of reference, the Menier Chocolate Factory can indeed be said to have influenced the development of early tall structures, just not in the way Giedion would have argued.

However, when looking at the outcome of the twentieth century, it is clear that Giedion's vision of the discrete frame rather than the heterogeneous combination of frame and infill became the norm. While it is beyond the scope of this essay to lay out the full mechanics of that reversal, consider the resulting multiple consolidations that characterize conventional contemporary construction: engineering practice—elastic analysis and plastic design[9]; material—industrialized steel and reinforced concrete; and systems—two-dimensional planar frames conducive to fast and easy calculation. All of these are now deeply embedded in building codes that developed as this consolidation took place. This implies that the challenge of building heterogeneously is more than the challenge of code approval but also requires a critical reevaluation of the embedded assumptions of both current building codes and the practice of engineering that are incompatible with most any form of heterogeneous construction. Fortunately, aspects of many of the buildings and prototypes considered in our current study provide paths to inform that reevaluation.

First consider Turkish hımış construction—represented in this study by *göz dolma*,[p 56] but also including the *çatkı* type.[10, Fig 2] *Çatkı* has been central to the expanded research by Randolph Langenbach on the performance of infill frame vernacular construction in Central Asia. In one case, he considers the performance of *çatkı* construction during the 1999 Turkey earthquake.[11] While examples of this traditional form suffered cosmetic damage, they were repairable, whereas many of their modern concrete frame neighbors collapsed.[Fig 3] Close inspection revealed that the subdivision of the wall into smaller panels in combination with low-strength mortar prevented the development of larger-scale global cracking, with the distributed cracking in the discrete masonry panels resulting in large energy dissipation. Structures of this type exhibit lateral capacity and ductile behavior as a system, even though they have low lateral strength and are composed of brittle materials.[12] While the performance of infill-frame interaction is difficult to quantify, the relatively higher performance of these structures in comparison to contemporary reinforced concrete frame buildings in these cases implies that the reliance of modern building codes solely on strength-based design is not without fault. The other insight related to this study is the importance of the relative performance of poor materials in combination that points to the potential for high performance from materials shunned in contemporary construction.[13]

Second, consider one of the contemporary examples in this study: the Kiln Tower[p 74] by Roger Boltshauser in collaboration with Jörg Habenberger of SEFORB Sarl. In this case, they are combining prefabricated earth blocks, CLT timber dividers between courses (which also serve as lifting surface and replace time-intensive earth infill between courses), heavy glulam timber at the roof level, and post-tensioned steel cables. The structure is made effectively monolithic through the compression enforced by the tensioned steel cables. Despite the variability of the earth blocks and the complex series of springs for each of the combined materials, the structure is cleanly rationalized through the measurement of tension in the cables. This singular measurement can be used over time to understand the creep of the system as a whole—both to understand its geometric change and to recalibrate the tension to maintain clear lateral behavior.[14] This implies that an unconventional combination of materials can be rationalized through the use of conventional means—in this case post-tensioning. Though unconventional, there are paths available within the current engineering design paradigm to enable the use of heterogeneous material assemblies.[15]

Third, consider the prototypesp 171 developed with students as part of this study, first for the exhibition *Sticks and Stones and* and later in a studio at UC Berkeley. These test constructions are conceived as diagrams to think through heterogeneity rather than true building components. Their conceit is that they are composed of materials found in any big-box construction store, and their aim is to produce examples falling into some of the formal categories of this study. It is not the intent of this project to capture and transplant the historical examples central to this study. Instead, the prototypes internalize the logics of their antecedents and use that logic as the starting point for the empirical study of systems and material combinations of their own. To illustrate the extreme heterogeneity of the components beyond simple material difference in the prototypes from *Sticks and Stones and*, we separately mapped the density, strength, and stiffness of the component materials.Fig 4 These prototypes are a means of translating heterogeneous ideas and applying them empirically for evaluation both visually and analytically. They represent a last aspect of the larger question of how to engage the limits of contemporary conventions by purposefully sidestepping them on the one hand while providing a locus for collaborative interdisciplinary engagement and experiment on the other. These prototypes are heterogeneous in both material and function, with the latter implying that there can be in these cases no separation of function (by discipline) into separate layers that forces an engagement and friction that is vital to spur creative discovery.

These three examples taken in the aggregate provide vectors to critique the conventional limits of design frameworks, materials, and systems that characterize the relative uniformity of contemporary construction. They illustrate the values of alternative design paradigms, the unconventional use of conventional means, and truly experimental construction. While none of these examples provides a single solution to how to build heterogeneously, it is only a matter of time until the remixing of the principles in each case will lead to one of many solutions that does.

1 See Kent Harries, Lola Ben-Alon, and Bhavna Sharma, “Codes and Standards Development for Nonconventional and Vernacular Materials,” in *Nonconventional and Vernacular Construction Materials: Characterisation, Properties and Applications*, ed. Kent Harries and Bhavna Sharma (Sawston, UK: Woodhead Publishing, 2019), 81–98.

2 See *2021 International Residential Code* (Washington, DC: International Code Council, 2020).

3 IRC section 104.11 states “The provisions of this code are not intended to prevent the installation of any material or to prohibit any design or method of construction not specifically prescribed by this code. The *building official* shall have the authority to approve an alternative material, design or method of construction upon application of the *owner* or the owner’s authorized agent. The *building official* shall first find that the proposed design is satisfactory and complies with the intent of the provisions of this code, and that the material, method of or work offered is, for the purpose intended, not less than the equivalent of that prescribed in this code in quality, strength, effectiveness, fire resistance, durability and safety.”

4 See Sigfried Giedion, *Space, Time and Architecture: The Growth of a New Tradition* (Cambridge, MA: Harvard University Press, 1959), 202–4.

5 See Randolph Langenbach, “Saga of the Half-Timbered Skyscraper: What Does Half-Timbered Construction Have to Do with the Chicago Frame,” *Proceedings of the Second International Congress on Construction History*, Cambridge University, UK, 2006.

6 See Randolph Langenbach, “From ‘Opus Craticium’ to the ‘Chicago Frame’: Earthquake-Resistant Traditional Construction,” *International Journal of Architectural Heritage* 1, no. 1 (2007): 29–59. Langenbach specifically references an article by Irving K. Pond in *Inland Architect and Builder* from 1883 in which the author reflects on timber frame and masonry infill structures observed in parts of Spain.

7 Langenbach, “Saga of the Half-Timbered Skyscraper,” 5–6.

8 Langenbach, “Saga of the Half-Timbered Skyscraper,” 7–8; and Langenbach, “From ‘Opus Craticium’ to the ‘Chicago Frame,’” 52.

9 For a comprehensive explanation of modern structural analysis and its development, see Jacques Heyman, *Structural Analysis: A Historical Approach* (Cambridge, UK: Cambridge University Press, 1998), section 10.5, 160–62.

10 *Çatkı* includes a roughly triangulated timber frame infilled with brick masonry and weak mortar.

11 Langenbach, “From ‘Opus Craticium’ to the ‘Chicago Frame,’” 36–41.

12 Langenbach, “From ‘Opus Craticium’ to the ‘Chicago Frame,’” 54–56.

13 An interesting note relative to this point is that while the theory of plastic design developed around the ductile materials of steel and reinforced concrete, it has been shown by Jacques Heyman to also apply to the brittle material of stone masonry relative to thrust line geometry rather than material strain; see Jacques Heyman, *The Stone Skeleton* (Cambridge, UK: Cambridge University Press, 1995), chapter 2. It is perhaps through the lens of a similar expansive use of plastic theory that additional means of rationalizing heterogeneous structures can be formed.

14 See Jörg Habenberger, “Prestressing Rammed Earth,” in *Pisé—Rammed Earth: Tradition and Potential*, ed. Roger Boltshauser, Cyril Veillon, and Nadja Maillard (Zurich: Triest Verlag, 2019), 261–65.

15 In my own experience, I have found that the use of soft materials in series with more rigid materials can play a similar role in rationalizing complex interactions. For Chicago Horizon, designed with my coauthors for the 2015 Chicago Architecture Biennial, we used reinforced neoprene pads captured by milled steel plates at the base of each timber column to allow for the accommodation of surface tolerance and provide a high coefficient of friction so that the structure did not require mechanical connection down to its foundation. More recently, at Guy Nordenson and Associates for the Museum of Contemporary Art San Diego, we applied reinforced neoprene pads between precast concrete units of a stair to rationalize the uncertainty of bearing concrete to concrete. The use of reinforced neoprene is common practice in bridge bearings and its well-understood properties from these applications make it possible to be translated into other uses.

Fig 1

Fig 2

Fig 3

Fig 1 Elevation of Menier Chocolate Factory from *Encyclopédie d'Architecture* 4–6 (1875–77)—a contemporary publication illustrating both the frame and the polychromatic brick.

Fig 2 Elevation of a *çatkı* type house. Note the similarity in construction (though rougher in appearance) to the Menier Chocolate Factory.

Fig 3 Two adjacent buildings in Turkey following the 1999 earthquake—one in the vernacular *çatkı* type with minor damage and one with a contemporary concrete frame that collapsed.

Fig 4 Property maps of three prototypes from our studies illustrating respectively the range of material strength, elasticity, and density.

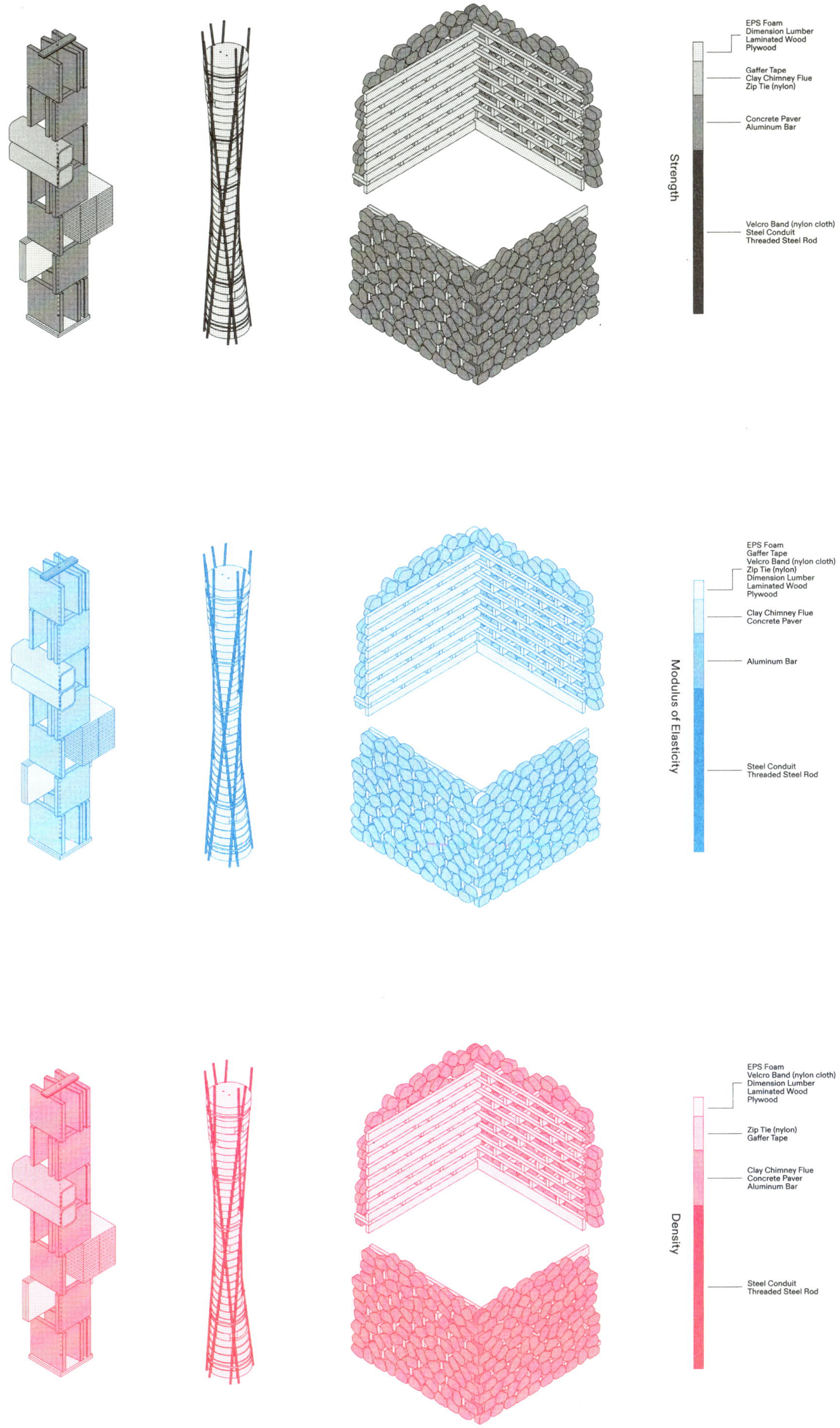

EPS Foam
Dimension Lumber
Laminated Wood
Plywood
Gaffer Tape
Clay Chimney Flue
Zip Tie (nylon)
Concrete Paver
Aluminum Bar
Velcro Band (nylon cloth)
Steel Conduit
Threaded Steel Rod
Strength
EPS Foam
Gaffer Tape
Velcro Band (nylon cloth)
Zip Tie (nylon)
Dimension Lumber
Laminated Wood
Plywood
Clay Chimney Flue
Concrete Paver
Aluminum Bar
Steel Conduit
Threaded Steel Rod
Modulus of Elasticity
EPS Foam
Velcro Band (nylon cloth)
Dimension Lumber
Laminated Wood
Plywood
Zip Tie (nylon)
Gaffer Tape
Clay Chimney Flue
Concrete Paver
Aluminum Bar
Steel Conduit
Threaded Steel Rod
Density

Fig 4

Photographs by
Naho Kubota

Name	COLLAGE 01
Location	Providence, Rhode Island
Year	2021
Team	Zak Leazer, Jordan Voogt, Wendy Zhuo
Construction	Concrete and Ceramic and Wood and Metal as Collage

This column comprises a single, mortarless stack of a variety of purchased and salvaged construction materials tied together by a steel tensioning rod. The stack is based on a 16-inch square module, with concrete pavers, plywood, mass timber, and clay elements swapped out at specific intervals. The post-tensioned rod combined with some softness in the materials contribute to the overall stability of the structure.

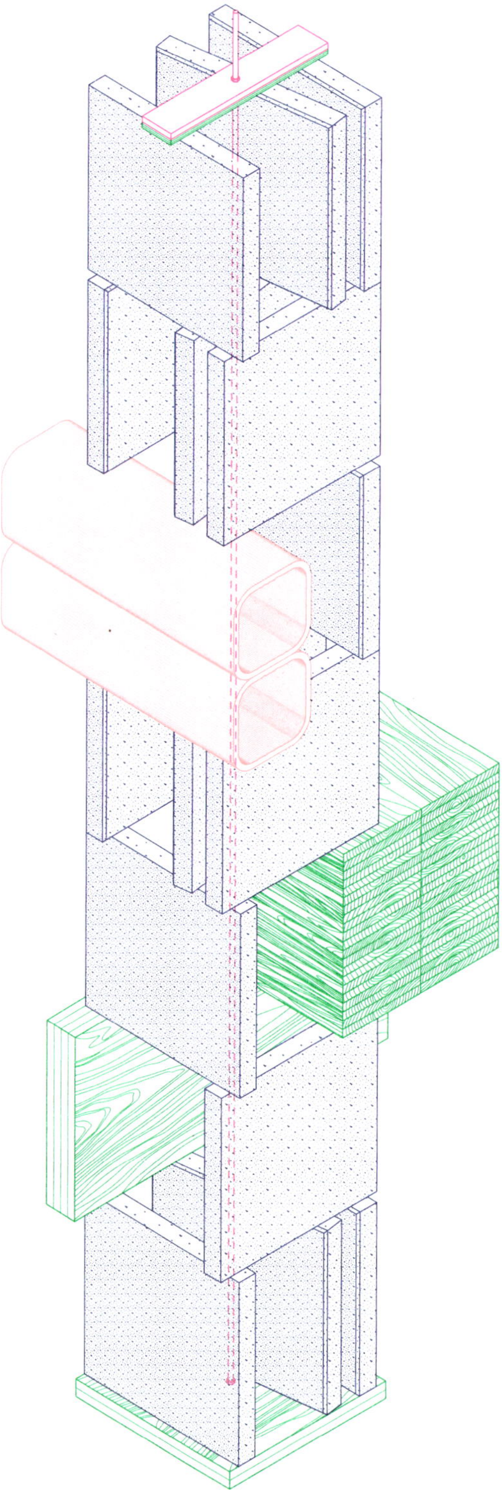

Concrete, Wood,
Ceramic, Metal

Name	WRAP 01
Location	Providence, Rhode Island
Year	2021
Team	Zak Leazer, Jordan Voogt, Wendy Zhuo
Construction	Metal and Other and Textile as Wrap

This column couples metal conduit with expanded polystyrene to create a strong but lightweight assembly capable of withstanding vertical compression loads. The conduit is crossed and tied together periodically to develop the full section. Straps around the assembly at regular intervals prevent buckling outward, while the foam provides resistance inward.

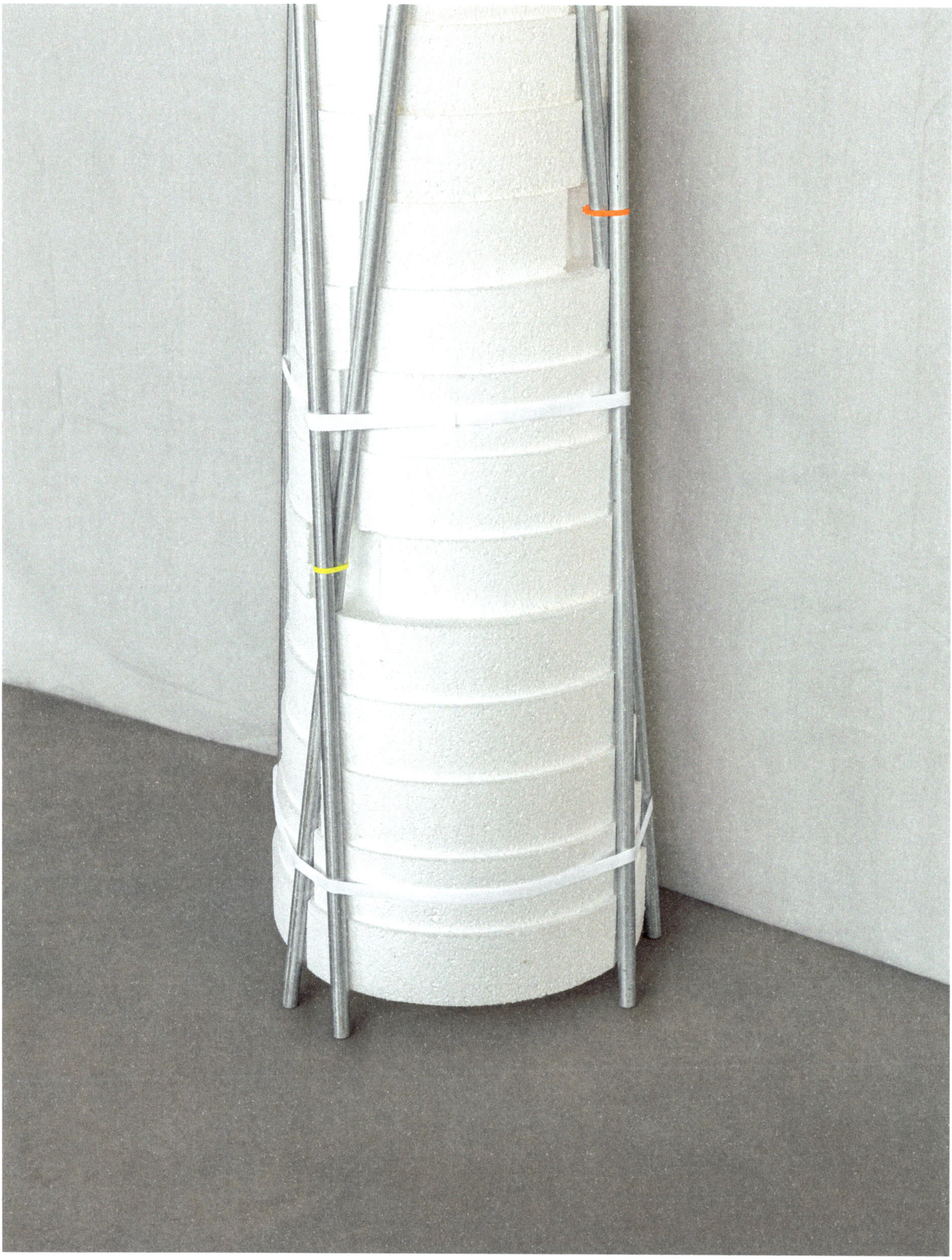

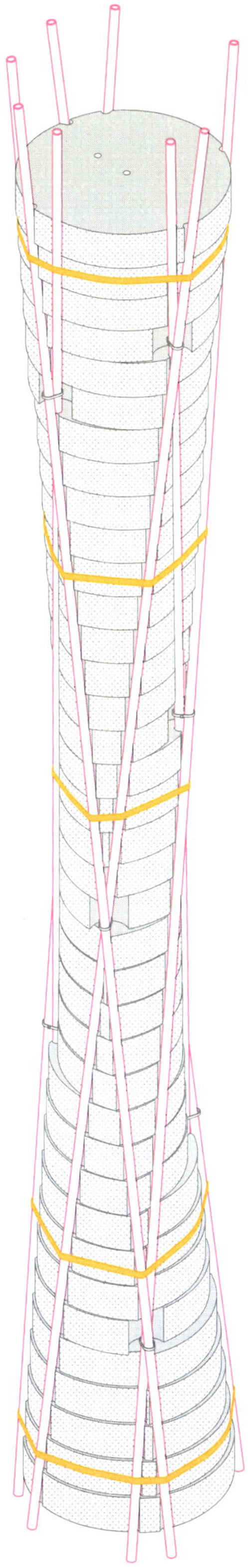

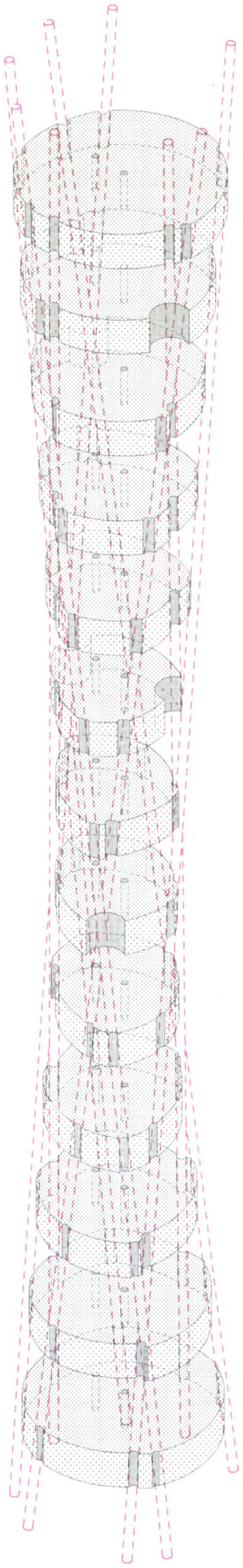

Name	SCAFFOLD 01
Location	Providence, Rhode Island
Year	2021
Team	Zak Leazer, Jordan Voogt, Wendy Zhuo
Construction	Concrete and Wood as Scaffold

This wall places two different materials—an irregular concrete paver and kiln-dried wood—into parallel, mortarless stacks. Alternate courses of the timber stack are rotated 90 degrees to form ties to the facing wythe. The two stacks scaffold one another and are built simultaneously to maintain stability. Textile straps between blocks at the corner provide friction and tie the blocks together using their own weight.

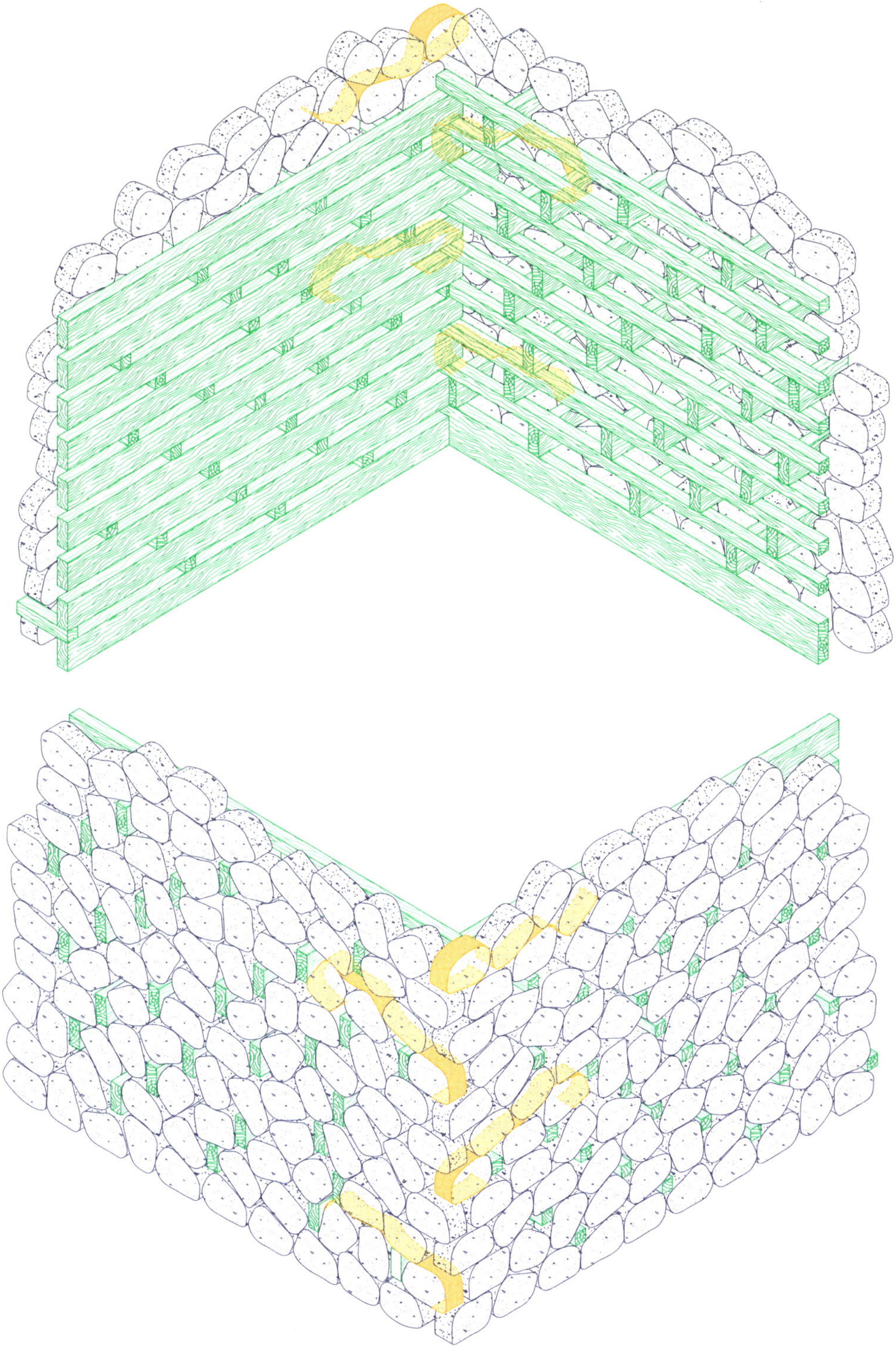

Name	SCAFFOLD 02
Location	Berkeley, California
Year	2023
Author	Zoe Frothinger, Ariana Wu
Construction	Wood and Concrete and Metal and Textile as Scaffold

This column makes use of a variety of found and salvaged construction materials sandwiched together to create a rotated stack of masonry-like compound elements. In the built version, wood and concrete are strapped to each other around a compressible textile core. The two materials are stacked individually—wood first, then concrete—then strapped together to provide a balance mechanism for each other. The construction sequence creates a self-scaffolding system, eliminating the need for cranes or other lifting equipment. Though built as a column, the sandwich-and-stack method can potentially be applied at a much larger scale, suggesting the creation of wall-like elements from large-scale demolition waste.

PACIFIC W
TECH

Name	FILL 01
Location	Berkeley, California
Year	2023
Author	Grisel Chavez, Oriana Lopez, Emily Tapia-Delgado
Construction	Ceramic and Wood as Fill

The wall design is an inversion of confined masonry, a construction system ubiquitous in developing countries around the world. In lieu of the typical concrete frame and masonry infill, this project combines a dry-stacked brick frame with timber bracing to provide lateral stability. The timber elements are milled to the dimension of the brick, enabling a seamless integration in the masonry coursing. The rotation of the brick from layer to layer produces a stepping effect at the corner, creating a clear distinction between this construction method and the seamlessness of other common trabeation systems.

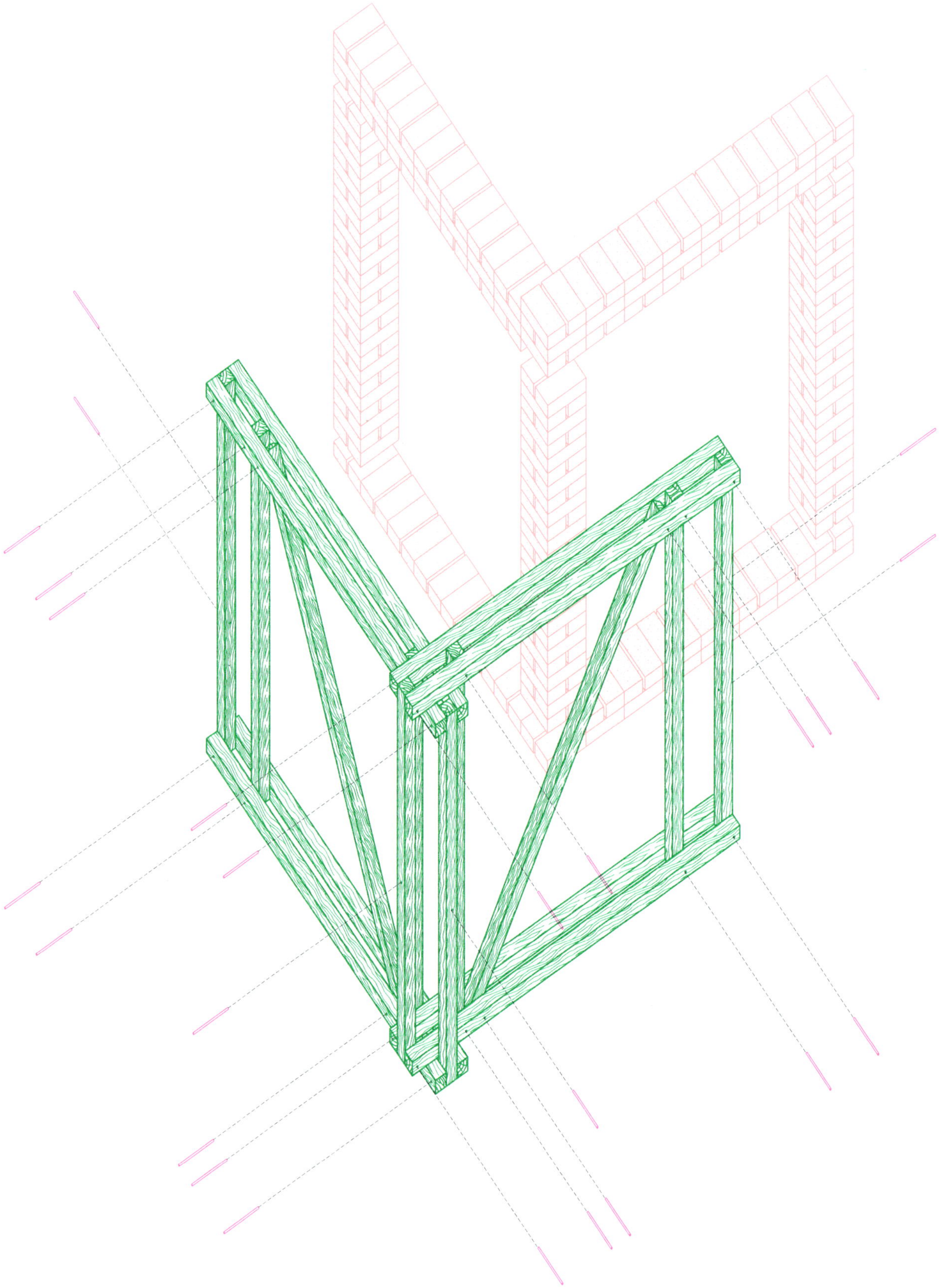

Name	WRAP 02
Location	Berkeley, California
Year	2023
Author	Mishelle Abusada, Grant Fisher, Jose Velasco
Construction	Wood and Other and Metal and Textile as Wrap

This column makes intelligent use of a variety of thin elements harvested from the construction waste stream and sorted into a self-assembling compound element. The materials are sorted into an ordered sequence with compressible foam rods at one end, wood at the other, and metal in the middle. The materials are woven into a mat using a method borrowed from bamboo mats used for making sushi rolls. Following this culinary precedent, the mat is then rolled up to create a lightweight and exceptionally strong column with the slender steel elements sandwiched between the foam core and wood outer layer, maximizing their compressive strength while containing inward and outward buckling. The nature of the weave and the strategic paring of the wood elements at the top and the bottom of the column create selective glimpses of each of the material layers.

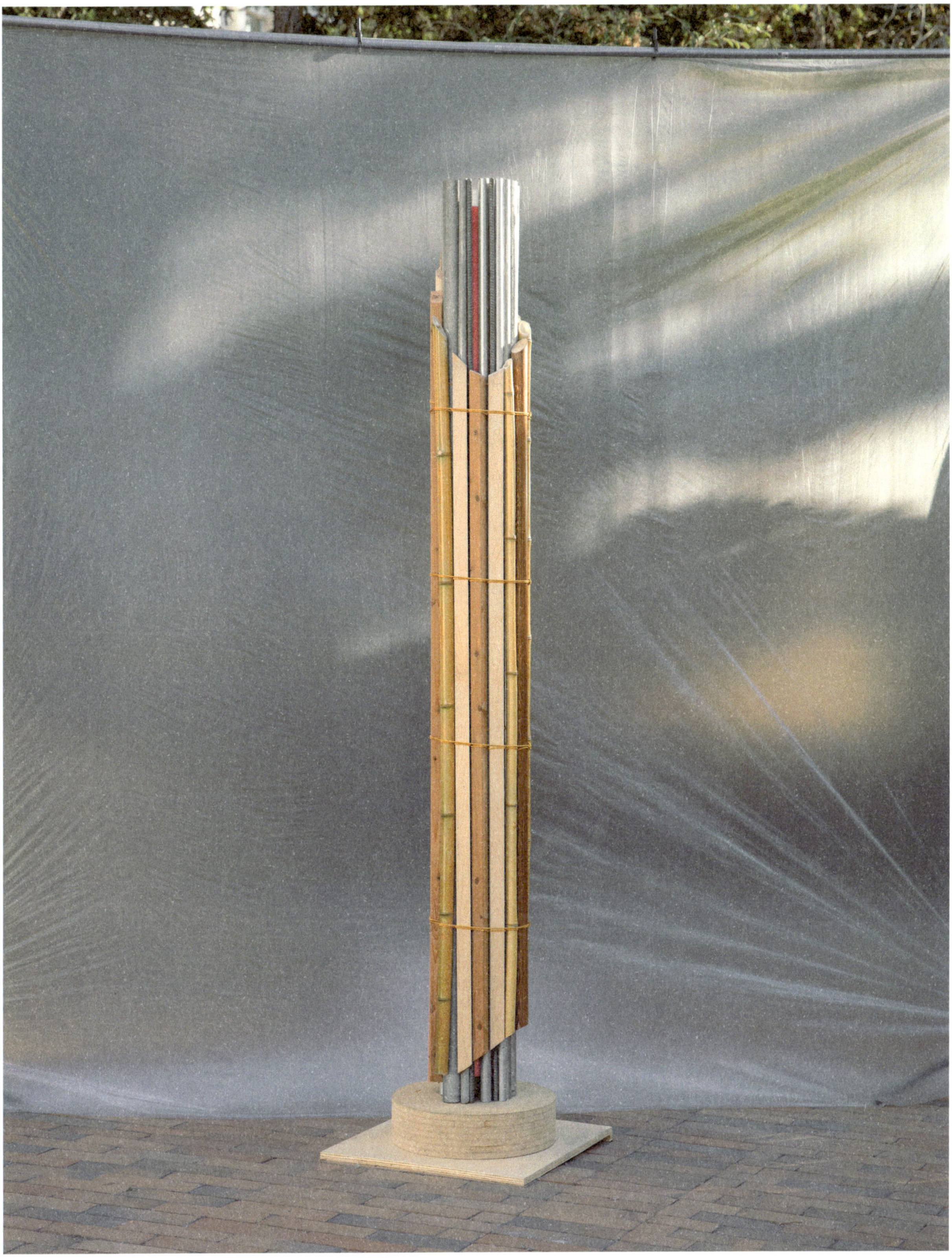

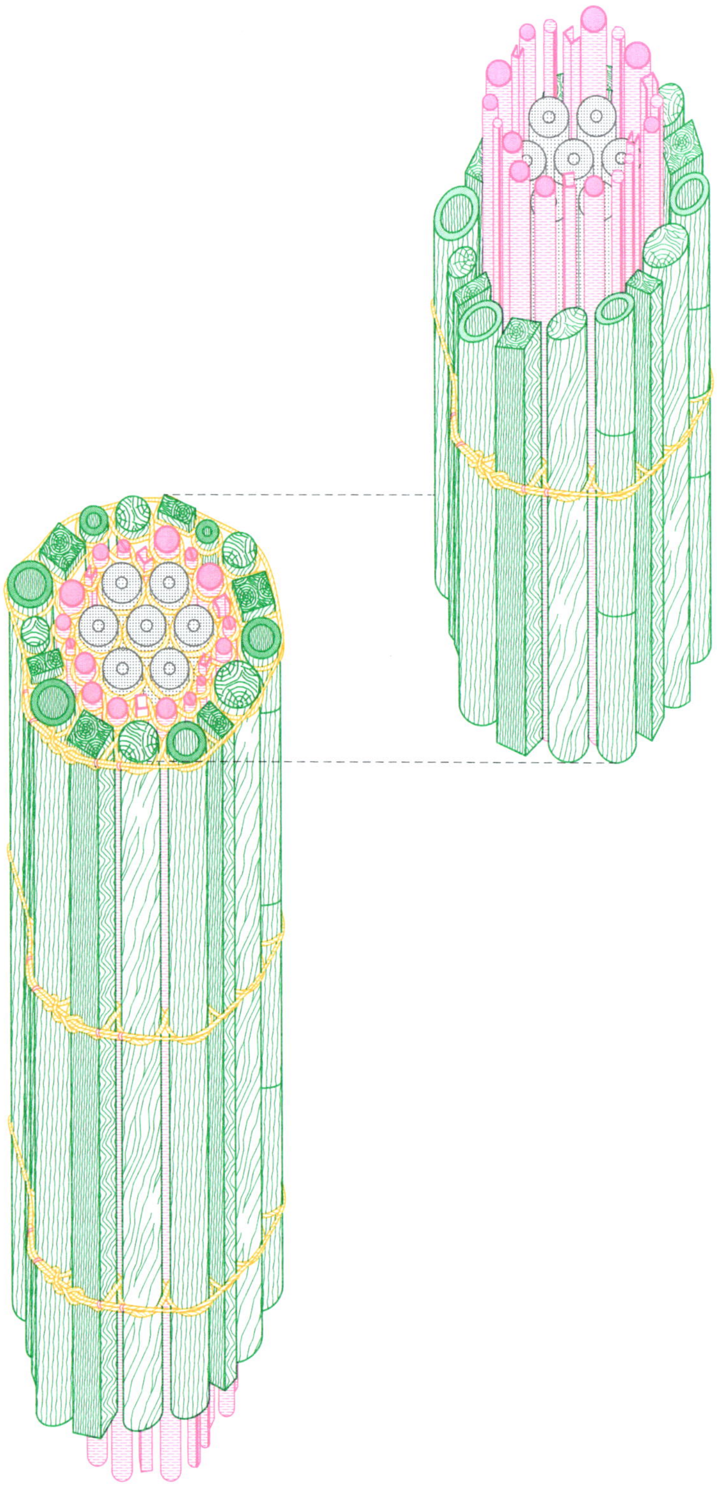

Wood, Other, Metal, Textile

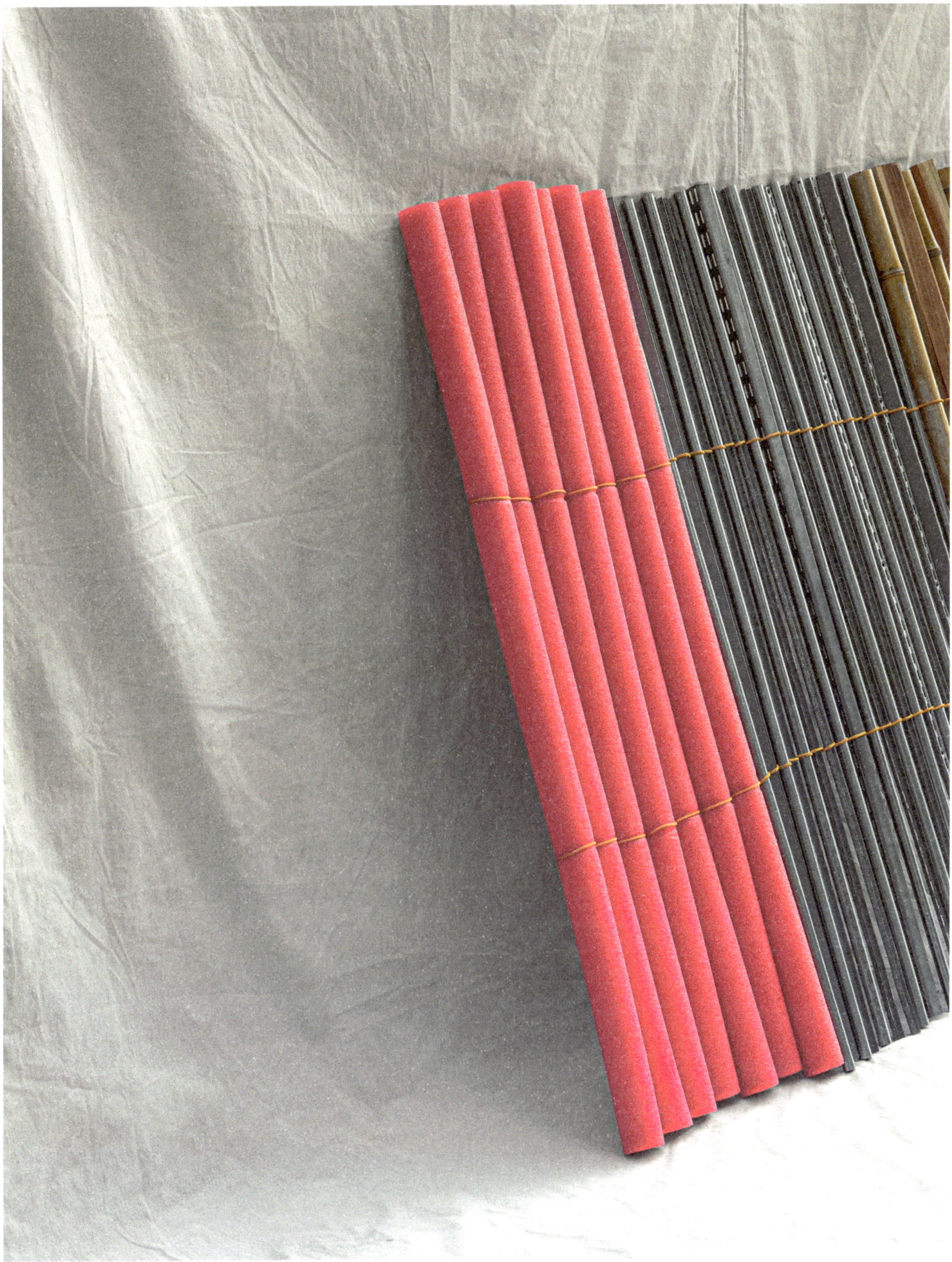

Name	SANDWICH 01
Location	Berkeley, California
Year	2023
Author	Maria Escalera, Ethan Lamping, Amanda Liang
Construction	Metal and Concrete as Sandwich, Earth and Wood as Sandwich

This highly rustic construction system combines masonry of various types with a lightweight, rigid vertical element that provides both a scaffolding mechanism and lateral bracing for the mortarless stack. The rotation of the masonry (which can vary depending on local availability) creates a corner condition with which the vertical element—typically a metal or wood rod—can mate to lock the masonry in place. Horizontal ties run diagonally across the masonry joints to sandwich the rods in place and to provide bracing. The repeatable geometry allows the creation of a variety of load-bearing structures, from pillars to bearing walls and cores. The simple elegance of the construction method suggests its usefulness in a range of scenarios, from DIY to professional.

 Concrete, Metal

Name	STACK 01
Location	Berkeley, California
Year	2023
Author	Samara Bhatia, Maria Hidalgo Rodas, Brant Zhao
Construction	Ceramic and Earth and Metal as Stack

The floor system is a play on both the New York Slab, an early concrete floor system that made use of thin, flexible reinforcing bars draped across tightly spaced iron bars, and the ubiquitous jack arch floor system of the late nineteenth century. In this case, a series of lightweight steel cables are draped between wood beams, with brick overlaid in a basket-weave pattern to support a "slab" of compacted earth. The system makes use of much lower carbon materials than its forebears while creating a simple in situ construction method that, like the jack arch, creates a rich and voluptuous ceiling condition that performs structurally, aesthetically, and acoustically.

See Also

Collage

Name Bayard Ewing Building
Location Providence, Rhode Island
Year 1848
Author Irving B. Hayes and Associates
Construction Brick, Metal, Wood as Collage

This former factory is typical of American Type III construction of the late nineteenth century. It combines multiple structural systems from floor to floor, including load-bearing masonry, cast-iron columns, and heavy timber joists, beams, and trusses.

Name Cappella Funeraria Molteni
Location Giussano, Italy
Year 1981
Author Aldo Rossi
Construction Brick, Concrete, Metal as Collage

This funerary chapel and tomb was constructed for the Molteni family in Giussano, Italy. Its brick walls are interrupted by insertions of concrete and steel elements.

Name Clemence–Irons House
Location Johnston, Rhode Island
Year 1691
Author Richard Clemence
Construction Stone, Wood as Collage

The Clemence-Irons House is a surviving example of an historical Rhode Island stone-ender. This small home has a stone wall and chimney that extend nearly the entire width of the building, while the rest of the home is timber-framed.

Name Domaine de la Terre, Îlot I
Location Villefontaine, France
Year 1985
Author Savigat, Perreau, Hamburger, and Munteanu
Construction Brick, Concrete, Earth as Collage

This model home in France was part of an earth-built social housing community. Loam was compacted in metal formwork before installation, with precast concrete and masonry framing window and door penetrations.

Name Frankish Castle
Location Paros, Greece
Year Since 13th century
Construction Stone, Wood as Collage

The castle is fabricated with intricate stonework that integrates elements of a fourth-century CE domed building atop the foundations of a temple from the sixth century BCE.

Name Gates of Nicaea
Location Iznik, Turkey
Year 1st century
Construction Brick, Stone as Collage

The gates of the ancient city are composed of the materials available at the time of construction, including spolia salvaged from nearby buildings collaged into a broader field of brick.

Name Habitat Initiative Cabo Delgado
Location Mozambique
Year 2010
Author Roswag Architekten
Construction Brick, Earth, Plants, Wood as Collage

This series of houses are characterized by a bamboo truss and palm leaf roof supported by walls that combine earth and brick according to their availability, strength, and water resistance.

Name House 1413
Location Ullastret, Spain
Year 2014
Author HARQUITECTES
Construction Concrete, Earth, Stone as Collage

The load-bearing walls of House 1413 are constructed using a combination of adobe and cyclopean techniques, which reuses stones from existing walls and mixes them with aggregate, limestone, and cement.

Name Jordi and Àfrica's Home
Location Mallorca, Spain
Year 2015
Author Ted'A Arquitectes
Construction Concrete, Stone as Collage

Upon a concrete base, old sandstone bricks taken from the site are repositioned alongside new sandstone bricks, which are used to frame openings in the facade. The various textures of the surface codify the materials according to whether they are original or new.

Name	JTB House
Location	Oud-Heverlee, Belgium
Year	2019
Author	BLAF Architecten
Construction	Ceramic, Wood, Plants as Collage

The walls of this house are made of extra-thick load-bearing bricks, with a timber frame infilled with dry lime hemp insulation on the inside face. In order to reduce complexity, the structural ceramic wall acts as both cladding and structure, and the dry lime hemp (which is naturally rot-resistant) eliminates the need for a vented cavity.

Name	Nam Dam Homestay and Community House
Location	Quản Bạ District, Vietnam
Year	2015
Author	1+1>2 Architects
Construction	Concrete, Earth, Wood as Collage

The foundation is constructed from local stone, the wall surface is coated with a mixture of earth and additives, and the structural frame is a combination of timber floor panels and reinforced concrete beams and columns with a timber roof.

Name	Spiller House
Location	Los Angeles, California
Year	1978
Author	Frank Gehry
Construction	Metal, Wood as Collage

The house is clad with a uniform corrugated metal sheathing, which pulls back to reveal light wood balloon frame construction underneath.

Name	Tibetan Farmhouse
Location	Phobjikha Valley, Bhutan
Year	Traditional
Construction	Earth, Stone, Wood as Collage

Tibetan farmhouses of this type consist of rammed clayey earth and inset timber. The cantilevered *rabsel* (the ornamental wood window frame) is inserted during the ramming process and is filled with bamboo wattle panels that are plastered with clay.

Name	Upper Lawn Pavilion
Location	Wiltshire, England
Year	1962
Author	Alison & Peter Smithson
Construction	Concrete, Stone, Wood as Collage

Responding to the ruins in situ, a wood frame structure is affixed to an old garden wall. An interior concrete beam preserves an old chimney by affixing it to the new structure.

Fill

Name	Cob House
Location	Fairfield, Iowa, USA
Year	2008
Author	Hap and Lin Mulleneaux
Construction	Earth, Plants, Wood as Fill

The word *cob* in this case means lumps that are formed of a mix of sand, clay, and straw that are then stacked and worked together while wet to form the walls of this house. The roof and openings are framed with timber.

Name	Foeth Dyes Workshop
Location	Germany
Year	1847
Construction	Brick, Wood as Fill

Forked timber columns organize the brick infill and work as natural braces for the structure in combination with the infill.

Name	Heuljip Earth Home
Location	Mokcheon, South Korea
Year	Traditional
Construction	Earth, Wood as Fill

This traditional type of house construction is typically round with earth wall construction stabilized with the addition of cordwood—intermittent timber rounds oriented through the wall with exposed ends. It is similar to cobwood construction in the West, but with fewer timbers.

Name	Çatkı House
Location	Safranbolu, Turkey
Year	Traditional
Construction	Earth, Stone, Wood as Fill

Çatkı is one type of *hımış* vernacular typical to the Black Sea region of Turkey that includes an infilled timber-frame construction system on a masonry podium. Stone or brick masonry with mortar are used to fill the triangulated voids in the timber frame.

Name	Hirose Family House
Location	Japan
Year	Since late 17th century
Construction	Earth, Stone, Wood as Fill

The timber grid of this house is infilled with clay, which serves to brace the grid laterally. The timber sits proud of the infill to prevent erosion. The technique dates from a period of timber scarcity in the region.

Name	House for an Art Collector
Location	Switzerland
Year	1986
Author	Herzog & de Meuron
Construction	Concrete, Wood as Fill

Precast concrete panels are separated by a timber frame in a wall system that appears similar to highway acoustic barriers.

Name	House in Noyers
Location	Noyers-sur-Serein, Yonne, France
Year	1600
Construction	Earth, Stone, Wood as Fill

The wall panels of this half-timbered home are filled with cob, stone, or torchis (a mixture of lime, straw, sand, and clay). When first built, the facade would have been clad with wood or slate shingles.

Name	House of the Opus Craticum
Location	Herculaneum, Italy
Year	60 CE
Construction	Brick, Stone, Wood as Fill

Piers of cemented brick support an upper story with a timber frame structure infilled with rubble.

Name	Korean Hanok
Location	Seoul, South Korea
Year	Since 14th century
Construction	Earth, Stone, Wood as Fill

Hanok houses typically consist of a timber frame infilled with earth floating above a stone base.

Name	Kuenzi Barn
Location	Wisconsin, USA
Year	1850
Construction	Brick, Wood as Fill

The barn is constructed with the German *Fachwerk* technique—a heavy timber frame forming roughly square panels is infilled with brick. The roof is framed with heavy timber and was originally covered with thatch.

Name	Muska House
Location	East Black Sea, Turkey
Year	Traditional
Construction	Stone, Wood as Fill

A masonry foundation is topped with a triangulated wood lattice frame that is infilled with stone.

Name	Saint-Simeon-de-Bressiuex Agriculture Building
Location	Saint-Siméon, Seine-et-Marne, France
Year	Traditional
Construction	Earth, Wood as Fill

This farm building is representative of a type, with a wide heavy timber frame infilled with rammed earth forming walls supporting a timber roof.

Name	Timber Weaver's Studio
Location	London, UK
Year	2019
Author	Practice Architecture
Construction	Concrete, Wood as Fill

The walls of this three-story building are made of a timber frame infilled with hempcrete. The hempcrete infill provides insulation and thermal mass while stiffening the timber frame.

Name	Wehlburg Artland Farmstead
Location	Lower Saxony, Germany
Year	1750
Construction	Brick, Wood as Fill

This Low German House uses a tightly gridded timber frame with brick infill and interior load-bearing stud walls.

Sandwich

Name	Atlántida Church
Location	Uruguay
Year	1960
Author	Eladio Dieste
Construction	Brick, Metal as Sandwich

This church uses steel-reinforced mortared brick in sinusoidal forms to create thin, undulating walls and long span roof vaults.

Name	Cyclopean House
Location	Brookline, MA
Year	2015
Author	Ensamble Studio
Construction	Other, Metal as Sandwich

This house is assembled with an ultra-light construction system consisting of light-gauge steel profiles sandwiching a foam core.

Name	Esperanza Dos
Location	Ecuador
Year	2011
Author	Al Borde
Construction	Plants, Stone, Wood as Sandwich

This collection of structures for a school in Ecuador uses wood and grasses woven together to form interlacing A-frame roofs.

Name	Otaniemi Chapel
Location	Espoo, Finland
Year	1957
Author	Heikki and Kaija Siren
Construction	Brick, Metal, Wood as Sandwich

Triangular trusses make use of paired timber in the chords and uprights sandwiching steel rods in the diagonals to create a lightweight, clear span over the chapel sanctuary.

Name	Ottoplatz Apartment and Office Building
Location	Switzerland
Year	1998
Author	Jüngling & Hagmann and Conzett Bronzini Partner AG
Construction	Concrete, Metal as Sandwich

This building has a post-tensioned precast concrete facade that is also the primary structure. The concrete panels partially overlap vertically to allow for post-tensioning cables to pass through and tie the patchwork of panels into an effective frame.

Name	Suoi Re Village Community House
Location	Luong Son, Hoa Binh Province, Vietnam
Year	2010
Author	1+1>2 Architects
Construction	Earth, Plants, Stone, Wood as Sandwich

The building uses a simple and economical structural system with the ground floor constructed from roughhewn stone and a bamboo ceiling. The second floor transitions to a rammed earth wall with heavy stone, bamboo frames, and palm leaf roof.

Name	Waigal Valley House
Location	Nuristan, Afghanistan
Year	Traditional
Construction	Stone, Wood as Sandwich

This construction method, called *piku-nakara,* is specific to this region of Afghanistan. Vertical timber poles are used to hold together a wall of stone masonry and lapped timber courses. The roof is framed with timber and covered with rammed earth.

Scaffold

Name	Button House
Location	Antalya, Turkey
Year	Since 18th century
Construction	Stone, Wood as Scaffold

Pairs of cedar beams alternate between rows of dry-stacked rubblestone masonry. Tie beams project from the wall at regular intervals, reinforcing the construction and creating a pattern across the facade.

Name	Cordwood Woodworking Shop
Location	Ontario, Canada
Year	1973
Author	John Gabriel Ötvös
Construction	Concrete, Wood as Scaffold

This building uses almost entirely recycled and local materials: cedar logs reclaimed from telephone poles, highway guardrails, and an abandoned log cabin. Sections of the logs are stacked and mortared like masonry and are protected by a roof with a deep overhang.

Name	Grand Parc Bordeaux
Location	Bordeaux, France
Year	2017
Author	Fréderic Druot and Lacaton & Vassal
Construction	Other, Concrete, Metal as Scaffold

A self-supporting structure was wrapped around the existing housing blocks to extend apartments and add new outdoor spaces. New components borrowed from greenhouse construction were prefabricated and assembled on-site, allowing residents to stay in their apartments during construction.

Name	Hale
Location	Hawaii
Year	Traditional
Construction	Plants, Wood as Scaffold

This Hawaiian vernacular uses a frame of timber poles—doubled as necessary for redundancy—that are joined and then lashed together using coconut fiber cordage. These buildings are typically open-walled with a thatch roof.

Name	House Rot Ellen Berg
Location	Braives, Belgium
Year	2011
Author	Architecten De Vylder Vinck Taillieu
Construction	Metal, Wood, Brick, Other as Scaffold

This house renovation uses industrial products—metal, glass, wood—in a contingent manner within the shell of an existing brick masonry building to create a house within a house.

Name	Larabanga Mosque
Location	Larabanga, Ghana
Year	1421
Author	Ayuba
Construction	Earth, Wood as Scaffold

The oldest mosque in Ghana was influenced by the architecture of western Mali of the same period, with palm timbers protruding from a mud brick-and-plaster exterior wall. The timbers tie the walls together laterally and serve as a scaffold for plaster repairs after each rainy season.

Name	Musgum Hut
Location	Cameroon
Year	Traditional
Construction	Earth, Plants as Scaffold

This traditional type is found in northern Cameroon. The exterior mud surface is applied against an interior surface of lashed thatch of reeds. The lashed reeds give the structure its pointed dome shape and the pattern of mud on the exterior creates steps to facilitate the seasonal maintenance of the dried mud.

Name	Nuristani House
Location	Nuristan, Afghanistan
Year	Traditional
Construction	Stone, Wood as Scaffold

This housing type is constructed of alternating rows of timber and stone with mud, with the interior wall frames piercing through for stability.

Name	Shabono
Location	Northern Amazon
Year	Traditional
Author	Yanomami Tribes
Construction	Plants, Wood as Scaffold

In this building type a communal void is circumscribed by a structure of wood and miscellaneous foliage that houses the entire tribe. The dwelling is periodically reconstructed to correspond with population shifts.

Name	Stove-Wood Barn
Location	Wisconsin, USA
Year	Since 19th century
Construction	Concrete, Wood as Scaffold

The stove-wood construction technique employs short pieces of rough-hewn wood stacked and mortared together as if they were bricks to form the main outer wall. The interior is framed with heavy timber with timber cladding boards above the main wall and on the roof.

Name	Teatro Oficina
Location	São Paulo, Brazil
Year	1984
Author	Lina Bo Bardi
Construction	Brick, Metal, Wood as Scaffold

The project inserts a permanent steel pipe scaffolding into the shell of a former brick masonry office building, bracing the existing wall while providing a seating rake and back-of-house space for an experimental theater.

Stack

Name Al-Abbas Mosque
Location Asnaf, Yemen
Year 1125
Author Muhammed ibn Abu l-Fathibn Arhab
Construction Brick, Stone, Wood as Stack

The lower walls of the Al-Abbas Mosque are stone with the upper portions constructed of mud brick. The roof is composed of finely carved wooden beams and is supported by six stone interior columns; four of them with stone capitals, two with stacked brick.

Name Bangali Thatched House
Location Bangladesh
Year Traditional
Construction Earth, Wood as Stack

Houses of this type use materials including reeds, long grass, rice straw, and jute sticks for the walls, with mud plastering on one or both sides. The roof is framed by bamboo and covered with thatch.

Name Corral Sculpture Studio
Location Vilches, Chile
Year 2015
Author Smiljan Radić
Construction Metal, Stone, Textile as Stack

A gantry crane supported on two pairs of triangulated columns and an existing corral wall respectively prop up and hold down a tensioned tarp roof.

Name Finishing Tower Sihlhölzli
Location Zurich, Switzerland
Year 2002
Author Boltshauser Architekten AG
Construction Concrete, Earth as Stack

This finishing tower is made of rammed earth walls with concrete floors and roof; lines of trass lime powder protect the rammed earth from erosion and contribute to a more abstract rendering of the rammed earth material, which would traditionally be protected by a large roof overhang.

Name Hexagonal Hogan
Location Canyon de Chelley, Arizona, USA
Year Traditional
Author Navajo Tribe
Construction Earth, Wood as Stack

The Navajo hogan is constructed of stacked logs with a corbelled roof vault. Earth covers the vault and fills gaps within the walls.

Name Icelandic Turf Home
Location Iceland
Year Traditional
Construction Earth, Stone, Wood as Stack

This construction technique was first brought to Iceland by the Vikings in the ninth century. Walls were typically constructed of stacked stone and peat blocks in herringbone patterns. Roofs were layered with timber, sod, and turf.

Name Kingaroy Peanut Silos
Location Australia
Year 1938
Author Thomas Robinson & Son
Construction Concrete, Metal, Wood as Stack

These concrete silos are topped with steel and timber sheds that house equipment for peanut processing.

Name Le Théâtre à L'exposition des Arts Décoratifs
Location France
Year 1925
Author Auguste Perret
Construction Concrete, Metal, Wood as Stack

This was a temporary theatre erected for the Exposition des Arts Décoratifs of 1925. Its concrete beams rest on heavy timber fluted columns. A steel armature spans the performance space, supporting white linen screens that filter daylight.

Name Log Tobacco Barn
Location Georgia, USA
Year 1977
Construction Wood as Stack

Tobacco barns from early twentieth-century Georgia are typically constructed by stacking logs longitudinally and vertically on each side of the barn and scribing notches where walls intersect to enable the locking of the corners.

Name	Meti Handmade School
Location	Bangladesh
Year	2005
Author	Anna Heringer and Eike Roswag
Construction	Earth, Textile, Wood as Stack

A bottom floor of cob construction lends support to a lighter second floor made of bamboo. Through these material differences, the first floor becomes cavernous whereas the bamboo level is airy, as further expressed by its textile ceiling.

Name	Otoprojects
Location	London, UK
Year	2013
Author	Assemble
Construction	Earth, Stone, Wood as Stack

Bags of compressed demolition rubble and earth are stacked on a disused site to create a monolithic volume, topped with a lightweight timber trussed roof.

Name	Primary School in Gando
Location	Burkina Faso
Year	2001
Author	Diébédo Francis Kéré
Construction	Brick, Concrete, Earth, Metal as Stack

The walls consist of hybrid clay/cement bricks protected from the rain by a large overhanging tin roof that is supported by steel bar trusses and concrete beams. A dry-stacked brick ceiling is introduced in between the classroom and the roof space to maximize ventilation.

Name	Ricola Storage Building
Location	Laufen, Switzerland
Year	1986
Author	Herzog & de Meuron
Construction	Concrete, Metal, Wood as Stack

This building is notable for its inversion of expected material hierarchies in contemporary construction. It is clad with cement paneling on a timber frame. Bonded timber posts are fixed to the back side of the framework to help support the weight of the cladding.

Name	Swiss Sound Pavilion
Location	Hannover, Germany
Year	2000
Author	Peter Zumthor and Jürg Conzett
Construction	Metal, Wood as Stack

Unseasoned wood beams are stacked and held in place by tension rods and steel springs that work in compression. The use of springs allows the dimension of the wood to change as it dries out, and for its reuse post-installation.

Name	Viking Longhouses
Location	Lofoton, Norway
Year	Since 8th to 11th centuries
Construction	Earth, Plants, Wood as Stack

These boat-shaped longhouses have two rows of wooden columns supporting the roof with a length up to 250 feet. Walls are made of earth, wooden planks, or clay, depending on availability of materials at the time of construction. The roof construction is composed of wood, turf, or thatch.

Wrap

Name	Bait Areesh
Location	United Arab Emirates
Year	Since 19th century
Construction	Plants, Wood as Wrap

These homes were the summer shelters of nomadic Bedouin tribes and are made of palm frond screens wrapped around a wooden frame of mangrove or palm poles topped with palm-thatched roofs.

Name	Center for Health Care and Social Promotion
Location	Burkina Faso
Year	2015
Author	Diébédo Francis Kéré
Construction	Brick, Earth, Stone as Wrap

The building contains a double-wall system with hollow concrete bricks on the outer envelope covered in clay and compressed clay bricks in the inner layer.

Name Cooling Tower
Location Germany
Year Since 20th century
Construction Concrete, Metal, Wood as Wrap

A sheeted holding tank is elevated on concrete piers supported by an exoskeletal steel framework.

Name Farman F.60 Goliath
Location France
Year 1919
Author Henri Farman
Construction Metal, Textile, Wood as Wrap

Similar to contemporaneous airplane construction, this prototype beam is made of canvas, wood, and wire as codependent structural elements.

Name Gamme
Location Nordland County, Norway
Year Traditional
Construction Earth, Wood as Wrap

This type of Sami housing uses an interior dome of curved poles to support an exterior of straw-reinforced earth blocks covered with sod which grows to match the surrounding landscape.

Name Hillhouse Box
Location Scotland
Year 2019
Author Carmody Groarke
Construction Concrete, Metal as Wrap

Working to preserve a heritage construction, this project wraps a new chainmail-clad steel structure around a historical building as a means of preservation. The result is one of heterogeneous layering in which each material may be identified via its temporality.

Name Kukje Gallery
Location Seoul, South Korea
Year 2012
Author SO-IL
Construction Concrete, Metal as Wrap

This gallery is made up of a series of site-cast concrete volumes that are wrapped in a mesh of stainless steel rings. The mesh lends the otherwise hard geometries of the building a fuzzy, light appearance.

Name New Life for a Peat Shed
Location Germany
Year 2015
Author Roswag Architekten, Guntram Jankowski
Construction Earth, Wood as Wrap

New walls of timber, cellulose, and earth are offset around an historical peat shed, marking their addition as a discrete building phase.

Name Oglala Lakota Tipi
Location South Dakota, USA
Year 1891
Author Lakota Tribe
Construction Leather, Wood as Wrap

For this demountable construction, concentric wood poles are held together by leather straps and covered with buffalo hides.

Name Rukari Palace
Location Nyanza, Rwanda
Year Since 18th century
Construction Plants, Wood as Wrap

The reconstruction of the palace of Rwanda features a large domed hall wrapped with plant materials and supported by slender wood columns.

Name Sanje Shigar House
Location Tibet
Year Traditional
Construction Earth, Stone, Wood as Wrap

This Tibetan type is typical in the Kongpo region. The walls are stone masonry bound by an earthen matrix. Less common slate stone is used only to frame openings. A timber grid frames the interior with timber internal partition walls and roof.

Name Stave Silo
Location USA
Year 1880 to 1950
Construction Metal, Wood as Wrap

Early grain silos were made of wood plank construction bound together with metal hoops, ties, and wires. The construction method fell out of popularity with the widening availability of clay-fired brick, concrete, and steel in the early 1900s.

Name	Villa Mairea
Location	Noormarkku, Finland
Year	1939
Author	Alvar Aalto
Construction	Textile, Wood as Wrap

Four wooden columns are separated by a wooden diaphragm at the center and bound together by a rope lashing. The binding prevents the individual poles from buckling under gravity loads.

Name	Yingzao Fashi Column
Location	China
Year	1368–1644
Construction	Metal, Textile, Wood as Wrap

A continuous column is assembled from multiple wood staves that are bound together by iron bands and enveloped in hemp fabric and mortar to achieve a smooth, monolithic appearance.

Name	Zulu Dwelling
Location	KwaZulu-Natal, South Africa
Year	1985 (reconstruction)
Construction	Leather, Textile, Wood as Wrap

The structure typically consists of local wood set into the ground and bent into a series of arches that interlink at the center of the house. The thatched roof made with grass and reed matting is held down with a net of grass ropes.

BIBLIOGRAPHY

ACHUM SHRINE

Geary, Christraud M. "Palace and Chieftaincy in Weh (North West Province, Cameroon)." In *Paideuma* 31 (1985): 183–201. Accessed July 27, 2022, http://www.jstor.org/stable/23076460.

Ritzenthaler, Robert E., and Pat Ritzenthaler. *Cameroons Village: An Ethnography of the Bafut*. Milwaukee: Milwaukee Public Museum, 1962.

BLACK TENT

Clarke, Gordon. "Tibet." In *Artsection*. Accessed July 27, 2023, http://artsection.org/tibet.html.

Faegre, Torvald. *Tents: Architecture of the Nomads*. Garden City, NY: Anchor Books, 1979.

Manderscheid, Angela. "The Black Tent in Its Easternmost Distribution: The Case of the Tibetan Plateau." In *Mountain Research and Development* 21, no. 2 (2001): 154–60.

Nomad Architecture. "Drokpa Nomads: Tibetan Black Tent." Uploaded December 8, 2017, https://www.youtube.com/watch?v=CAIMc9Zbg5I.

Schittich, Christian. "Yurts and Black Tents in the High Steppes of Asia." In *Vernacular Architecture: Atlas for Living Throughout the World*, edited by Christian Schittich, 166–71. Basel: Birkhäuser, 2021.

CATTLE SHED IN COSSIAT

Boltshauser, Roger, Nadja Maillard, Cyril Veillon, and Romain Anger. *Pisé—Rammed Earth: Tradition and Potential*. Zurich: Triest Verlag, 2019.

"François Cointeraux." In *Wikipedia*. Accessed April 14, 2023, https://en.wikipedia.org/w/index.php?title=Fran%C3%A7ois_Cointeraux&oldid=1044905790.

Royer, Claude. *L'architecture rurale française: Corpus des genres, des types et des variantes Lyonnais*. Paris: Berger-Levrault, 1979.

GÖZ DOLMA

Arun, Görün. "Evolution of Timber Construction in Turkey." In *Archi-Cultural Translations through the Silk Road, 2nd International Conference, Mukogawa Women's University, Nishinomiya, Japan, July 14–16, 2012, Proceedings* (2012), 225–30.

Bal, İhsan E., F. Gülten Gülay, and Meltem Vatan. "Structural and Non-structural Damages of Turkish Vernacular Timber Frame Structures." Paper presented at the 9th International Symposium on Conservation of Monuments in Mediterranean Basin, Middle East Technical University, Ankara, Turkey, June 2014.

Batur, Afife, and Öymen Gür Şengül. *Doğu Karadeniz'de kırsal mimari / Rural Architecture in the Eastern Black Sea Region*. Istanbul: Milli Reasürans, 2005.

Gülkan, Polat, and Randolph Langenbach. "Traditional Timber-Laced Masonry Construction in Turkey Known as Himiş." In *Masonry Construction in Active Seismic Regions*, edited by Rajesh Rupakhety and Dipendra Gautam, 61–97. Sawston, UK: Woodhead Publishing, 2021.

İlhan, Tolga. "Construction Techniques and Building Processes of Traditional Timber Frame Arhavi Houses." Diss., Middle East Technical University, Ankara, 2019.

GREAT MOSQUE OF DJENNÉ

Alatalo, Erika. "Living Heritage and Earth Architecture of Djenné." In *Field Study of the World*, November 22, 2018, https://www.fieldstudyoftheworld.com/living-heritage-earth-architecture-djenne/.

ArchEyes. "The Great Mosque of Djenné: An Exploration of Traditional Mud Architecture." Accessed April 27, 2016, https://archeyes.com/great-mud-architecture-mali-dogon-culture/.

Wondermondo. "Great Mosque of Djenne (Djenné)." Accessed July 27, 2023, https://www.wondermondo.com/great-mosque-of-djenne/.

Yeung, Peter. "The Massive Mosque Built Once a Year." In *BBC*, August 2, 2019, https://www.bbc.com/travel/article/20190801-the-massive-mosque-built-once-a-year.

Zamani Project. "Great Mosque of Djenné & Djingareyber Mosque." In *Black Monuments Matter*. Accessed July 27, 2023, https://black-monuments-matter.zamaniproject.org/site-mali-mosques.html.

KATH-KHUNI

Ahuja, Rita, and Aditya Rahul. "Analysing Sustainability Issues Related to Various Construction Techniques Practiced in Himalayas." In *International Journal of Research* 1, no. 5 (June 2014): 213–18.

Dave, Bharat, Jay Thakkar, and Mansi Shah. *Prathaa: Kath-khuni Architecture of Himachal Pradesh*. CEPT University, India: SID Research Cell, 2013.

Dhandapany, D., and Arun Menon. "Seismic Behavior of Timber-Laced Masonry Structures in the Himalayan Belt." In *Structural Analysis of Historical Constructions*, edited by Rafael Aguilar, Daniel Torrealva, Susana Moreira, Miguel A. Pando, and Luis F. Ramos, 605–13. Cham: Springer International Publishing, 2019.

Raghuvanshy, Garima. "The Story of Kath Kuni: Preserving Vernacular Architecture in the Himalaya." In *Himal Southasian*, September 22, 2021, https://www.himalmag.com/the-story-of-kath-kuni-preserving-vernacular-architecture-himalaya-2021/.

Shah, Mansi, and Jay Thakkar. "The Himalayan Vernacular: Kath-khuni Architecture." In *Sahapedia*, September 19, 2018, https://www.sahapedia.org/the-himalayan-vernacular-kath-khuni-architecture.

KILN TOWER FOR THE BRICKWORKS MUSEUM

Baumann, Jason. "Kiln Tower 2021—Studio Boltshauser—Future Now Pavilion—Insight Architecture," August 28, 2021, https://www.youtube.com/watch?v=OVHwUV6NLEM.

Boltshauser Architekten. "Kiln Tower for the Brickworks Museum." Accessed July 27, 2023, https://boltshauser.info/en/projekt/ofenturm-fuer-das-ziegelei-museum/.

Boltshauser, Roger, Nadja Maillard, Cyril Veillon, and Romain Anger, eds. *Pisé—Rammed Earth: Tradition and Potential*. Zurich: Triest Verlag, 2019.

Rauch, Martin, Otto Kapfinger, and Marko Sauer. *Martin Rauch: Refined Earth: Construction & Design with Rammed Earth*, edited by Otto Kapfinger. 3rd rev. and exp. ed. Munich: DETAIL Business Information GmbH, 2022.

MENIER CHOCOLATE FACTORY

Calliat, Victor, and Adolphe Lance. *Encyclopédie d'architecture*, vol. 3. Paris: Bance, 1853.

Giedion, Sigfried. *Building in France, Building in Iron, Building in Ferroconcrete*. Translated by J. Duncan Berry, with an introduction by Sokratis Georgiadis. Santa Monica, CA: Getty Center for the History of Art and the Humanities, 1995.

Giedion, Sigfried. *Space, Time and Architecture*. Cambridge, MA: Harvard University Press, 2009.

Picon, Antoine, and Centre Georges Pompidou. *L'art de l'ingénieur: constructeur, entrepreneur, inventeur*. Exh. cat. Centre Georges Pompidou. Paris: Le Moniteur, 1997.

Saulnier, Jules. "Maquette Du Moulin de La Chocolaterie Menier, Noisiel," 1871–72. Cité de l'architecture et du patrimoine. Accessed July 27, 2023, https://www.citedelarchitecture.fr/fr/collection/la-maquette-analytique-un-heritage/maquette-du-moulin-de-la-chocolaterie-menier-noisiel.

NATIONAL MUSEUM OF ROMAN ART

Allen, Stan. "Drawing with Rafael Moneo, Madrid 1984." In *Drawing Matter*, March 14, 2019, https://drawingmatter.org/stan-allen-on-drawing-with-rafael-moneo-madrid-1984/.

Deplazes, Andrea. "Prefabrication and opus caementitium," in *Constructing Architecture: Materials, Processes, Structures: A Handbook*. Basel: Birkhäuser, 2005.

Langdon, David. “AD Classics: National Museum of Roman Art / Rafael Moneo.” In *ArchDaily*. Accessed October 1, 2018, https://www.archdaily.com/625552/ad-classics-national-museum-of-roman-art-rafael-moneo.

Moneo, José Rafael, et al. *Rafael Moneo: Remarks on 21 Works*. New York: Monacelli Press, 2010.

Moneo, Rafael, Fernando Márquez Cecilia, and Richard Levene. *Rafael Moneo: 1967–2004*. El Escorial (Madrid): El Croquis, 2004.

SCHOOL IN ORSONNENS

“2014 | School in Orsonnens: TEd’A Arquitectes.” In *TEd’A Arquitectes*. Accessed July 27, 2023, http://www.tedaarquitectes.com/english/index.php?/projects/-2014-school-in-orsonnens/.

Fernando Márquez, Cecilia, and Richard Levene. *TEd’A Arquitectes 2010–2018: Materia de Juego: Material in Play*. El Escorial (Madrid): El Croquis, 2018.

Ramis, Toni, Tomeu Mateu, Margherita Lurani, and Teresa Piferrer. “TEd’A Arquitectes, Rapin Saiz Architectes, School in Orsonnens.” In *Divisare*, June 1, 2018, https://divisare.com/projects/387170-ted-a-arquitectes-rapin-saiz-architectes-luis-diaz-diaz-school-in-orsonnens.

“School in Orsonnens.” In *DETAIL Inspiration*, 2020. Accessed February 27, 2024, https://inspiration.detail.de/school-in-orsonnens-114650.html?lang=en.

SENECA BARK LONGHOUSE

“Bark Longhouse.” In *Ganondagan*. Accessed July 27, 2023, https://www.ganondagan.org/Visit/Bark-Longhouse.

“Ganondagan State Historic Site.” In *Wikipedia*. Accessed July 27, 2023, https://en.wikipedia.org/w/index.php?title=Ganondagan_State_Historic_Site&oldid=1145908640.

“Parks, Recreation and Historic Preservation.” In *Ganondagan State Historic Site*. Accessed July 27, 2023, https://parks.ny.gov/historic-sites/26/details.aspx.

TOMIOKA SILK MILL

Nishio, Toshikazu, Shinya Tsukada, Tetsuo Morita, and Akira Yuzawa. “A Study on Town Planning and Evaluation of Industrial Heritage Tomioka Silk Mill.” In *Journal of Architecture and Planning* 79, no. 705 (2014): 2507–16.

Shogaki, Takaharu, Yukiyasu Fujii, and Yuhta Nakagawara. “Deformation of the Foundation and Structure of Tomioka Silk Mill’s East Cocoon Warehouse.” In *Soils and Foundations* 59, no. 4 (2019): 789–800.

“The World Heritage Tomioka Silk Mill: A Modern Factory with Japanese Flavor.” In *Nippon*, July 1, 2023, https://www.nippon.com/en/guide-to-japan/gu900122/the-world-heritage-tomioka-silk-mill-a-modern-factory-with-japanese-flavor.html.

“Tomioka Silk Mill and Related Sites.” In *UNESCO World Heritage Centre*. Accessed July 27, 2023, https://whc.unesco.org/en/list/1449.

“UNESCO Listing for Tomioka Silk Mill: Meiji-Era Structure a Symbol of Japan’s Modernization.” In *Nippon*, May 1, 2014, https://www.nippon.com/en/behind/l00061/.

TULOU

Colafranceschi, Elena, Elisabetta Pallottino, and Paola Porretta. “Tulou: The Rammed Earth Dwellings of Fujian (China). Functional, Typological and Constructive Features.” In *ISPRS — International Archives of the Photogrammetry, Remote Sensing and Spatial Information Sciences* (2020).

Guo, Qinghua. “The Tulou Buildings in Fujian.” In *Vernacular Architecture: Atlas for Living Throughout the World*, edited by Christian Schittich, 172–81. Basel: Birkhäuser, 2019.

Huang, Hanmin. *Fujian’s Tulou: A Treasure of Chinese Traditional Civilian Residence.* Singapore: Springer Singapore, 2009.

Senses Atlas. “Fujian Tulou, the Hakka Walled Villages.” In *Senses Atlas*. Accessed November 24, 2020, https://www.sensesatlas.com/territory/fujian-tulou-the-hakka-walled-villages/.

UMGEBINDEHAUS

Richter, Kerstin. “Umgebinde Houses in the Border Triangle Between the Czech Republic, Germany and Poland.” In *Vernacular Architecture: Atlas for Living Throughout the World*, edited by Christian Schittich, 66–73. Basel: Birkhäuser, 2019.

Thomas, Worbs, and Liane Vogel. “Informationszentrum Umgebindehaus.” In *Informationszentrum Umgebindehaus*. Accessed July 27, 2023, https://umgebindehaus.hszg.de/.

“Upper Lusatian house.” In *Wikipedia*. Accessed July 27, 2023, https://en.wikipedia.org/w/index.php?title=Upper_Lusatian_house&oldid=1155618643.

YURT

Drew, Philip. *Tensile Architecture*. London: Granada, 1979.

Faegre, Torvald. *Tents: Architecture of the Nomads*. Garden City, NY: Anchor Books, 1979.

Schittich, Christian. “Yurts and Black Tents in the High Steppes of Asia.” In *Vernacular Architecture: Atlas for Living Throughout the World*, edited by Christian Schittich, 160–66. Basel: Birkhäuser, 2019.

“Yurt.” In *Wikipedia*. Accessed July 27, 2023, https://en.wikipedia.org/w/index.php?title=Yurt&oldid=1164841028.

IMAGE CREDITS

Sigfried Giedion / Getty Publications: p 31 Fig 1
© BuildingScience.com: p 31 Fig 2
© Fabian - stock.adobe.com: pp 38, 39, 41
Gordon Clarke: pp 44, 45, 47
Philip Heckhausen: pp 50, 51, 52
Mustafa Reşat Sümerkan: pp 56, 57, 59
Hamdia Traoré: pp 62, 63, 65
Mansi Shah and Jay Thakkar: pp 68, 69, 71
Kuster Frey: pp 74, 75 (r)
© Sandro Livio Straube: pp 75 (l), 77
Wikimedia Commons / Myrabella: pp 80, 81 (r)
Flickr / besopha: pp 81 (l), 83
OTTO / Michael Moran: pp 86, 89
Rafael Moneo Arquitecto: p 87
Luis Diaz Diaz: pp 92, 93 (l), 95
TEd'A Arquitectes: p 93 (r)
Philip Scalia / Alamy Stock Photo: p 99 (r)
Kei Takanami: pp 104, 105, 107
© Sakawduen — stock.adobe.com: p 110
© rabbit75_fot — stock.adobe.com: pp 111 (l), 113
© Xin Wang — stock.adobe.com: p 111 (r)
© Robert Rieger. Restoration by Atelier ST: pp 116, 117 (l), 119
Wikimedia Commons / Matěj Baťha: p 117 (r)
Rhett Butler: pp 122, 125
Wikimedia Commons / Tkn20: p 123 (l)
Jeremy Graham / Alamy Stock Photo: p 123 (r)
Michael F. Ashby. Reprinted from *Materials Selection in Mechanical Design*, Michael F. Ashby, "Materials selection charts," p 37, © 2011, with permission from Elsevier: p 137 Fig 1
Ludwig Mies van der Rohe © 2024 Artists Rights Society (ARS), New York / VG Bild-Kunst, Bonn. Digital Image © The Museum of Modern Art/ Licensed by SCALA / Art Resource, NY: p 137 Fig 2
Joachim Brohm © 2024 Artists Rights Society (ARS), New York / VG Bild-Kunst, Bonn: p 137 Figs 3, 4
Eugène-Emmanuel Viollet-le-Duc: p 145 Figs 1, 2
KARAMUK KUO: p 145 Figs 3, 4
John Nicolais Collection. The Architectural Archives, University of Pennsylvania © Frank O. Gehry / Getty Research Institute, Los Angeles (2017.M.66): p 153 Fig 1
© Robert Rauschenberg Foundation. Digital Image: The Art Institute of Chicago / Art Resource, NY : p 153 Fig 2
Lais Myrrha, Dois pesos, duas medidas (2016–17). In *Live Uncertainty: An exhibition after the 32nd Bienal de São Paulo*. Fundação de Serralves—Museu de Arte Contemporânea, Porto, Portugal. June 30, 2017–May 1, 2018. Photo © Filipe Braga: p 153 Fig 3
Sebastian Schels: p 153 Fig 4
Courtesy Albuquerque & Sons: p 158
Ajay Manthripragada: pp 159, 160, 161
Encyclopédie d'Architecture 4–6 (1875–77): p 168 Fig 1
Randolph Langenbach: p 168 Fig 2
Adem Dogangun: p 168 Fig 3
© Naho Kubota: pp 173, 175, 177–8, 181–2, 185–6, 189, 191, 193, 195, 197, 199, 201–2, 204–5

BIOGRAPHIES

Aaron Forrest is an architect and co-founder of the architecture office Ultramoderne. He is currently an associate adjunct professor of architecture at the University of California at Berkeley.

Andrew LeClair is a graphic designer based in New York. He currently teaches at Parsons School of Design.

Naho Kubota was born in Iida, Japan, and is an artist and photographer based in New York.

Jeannette Kuo is a founding partner of KARAMUK KUO Architects based in Zurich and professor of architecture and construction at TU Munich.

Ajay Manthripragada is a practicing architect with current projects in California and India. He is an assistant professor of architecture at the University of California at Berkeley.

Brett Schneider is an engineer based in Houston, Texas, and Providence, Rhode Island. He is an associate professor of architecture at Rice University, and a senior associate at Guy Nordenson and Associates Structural Engineers in New York.

Jesús Vassallo is a practicing architect and writer based in Houston, Texas. He is currently an associate professor of architecture at Rice University.

Yasmin Vobis is an architect and assistant professor of architecture at the University of California at Berkeley. She is co-founder of the architecture office Ultramoderne.

Heterogeneous Constructions has followed a winding path on its way from project to publication. The research began as the extension of collaborations in practice between the authors. It initially took shape as a spreadsheet and culminated in an exhibition at the Rhode Island School of Design, with subsequent incarnations as seminars at Harvard GSD and a design studio at UC Berkeley. The students, colleagues, and collaborators who have influenced the project in various ways are too numerous to list here, but we will make an attempt regardless:

The dozens of students who contributed their efforts and design expertise to the drawings and the prototypes at the center of this book. Without their inspiring and endless creativity, nothing would have been possible.

Andrew LeClair deserves recognition for his tireless efforts on the design of the book as well as the initial exhibition. The design of the formats for presenting the work was as integral to the project as any drawing or text.

Jesús Vassallo, Jeannette Kuo, and Ajay Manthripragada for the essays they contributed to this book and for the many conversations that we have been having with each of them since long before this project ever existed. We are honored to count such brilliant thinkers as friends.

Marlon de Azambuja, Roger Boltshauser, John Ochsendorf, and Jesús Vassallo all took part in formative public conversations on the topic prior to the initial exhibition.

Katharina Kulke and Anja Haering at Birkhäuser provided expert guidance on the content and production of the book to make it what it is today.

The case study documentation in particular was based on in-depth field research by many others. We are grateful for their untold work that went into recording these building practices in minute detail, and for the input of those who took the time to review our own drawings for accuracy. We are also grateful to the contemporary architects who provided us with additional information about their buildings to be included here.

UC Berkeley, Harvard GSD, and RISD, who made available resources to make the project reality. Many individuals are noted in the credits, but Lisa Iwamoto deserves particular recognition for laying the groundwork for the studio where many of the student prototypes were realized.

Guy Nordenson, who brought us together as collaborators for the first time in New York; and Laura Briggs, who did the same again many years later at RISD.

Finally, and most importantly, our families, who have supported us along the way.

Our deepest thanks to all.

CREDITS

Project Leads
Aaron Forrest
Brett Schneider
Yasmin Vobis

Graphic Design
Andrew LeClair

Prototypes Photography
Naho Kubota

Research Team
Lara Avram, Pericles de la Torre Bien, Evelyn Ehgotz, Zoe Frothinger, Nicola Ho, Zak Leazer, Diandra Rendradjaja, Kei Takanami, Jordan Voogt, Ariana Wu, Wendy Zhuo

UC Berkeley Design Studio Fall 2023
Mishelle Abusada, Samara Bhatia, Grisel Chavez Diaz, Maria Escalera, Grant Fisher, Zoe Frothinger, Maria Hidalgo Rodas, Ethan Lamping, Amanda Liang, Oriana Lopez, Emily Tapia-Delgado, Jose Velasco Ochoa, Ariana Wu, Brant Zhao

Institutional Support
RISD Department of Architecture
Harvard Graduate School of Design
UC Berkeley College of Environmental Design

Fabrication Support
RISD: David Dilks, James Dean
Harvard GSD: Rachel Stefania, Jon Gregurick
UC Berkeley: Semar Prom, Mike Bond

Funding
RISD Architecture Design Research Seed Fund
Harvard GSD Faculty Research Fund
Harvard GSD Dean's Jr. Faculty Research Grant
UC Berkeley Faculty Research Fund
UC Berkeley Book Subvention Program
UC Berkeley Conway Fund

All texts, drawings, and images are by the authors, unless otherwise indicated. All drawings of and data regarding historical buildings are based on available documentation and may not be fully accurate. Any known errata will be corrected in future editions.

Project Management
Katharina Kulke, Alexander Felix

Production
Anja Haering

Copy Editing
Keonaona Peterson

Reproduction
Repromayer Medienproduktion, Reutlingen

Printing
Grafisches Centrum Cuno, Calbe

Fonts
ABC Diatype, ABC Synt

Paper
Munken Lynx, Peydur Lissé

Library of Congress Control Number: 2024941061

Bibliographic information published by the German National Library

The German National Library lists this publication in the Deutsche Nationalbibliografie; detailed bibliographic data are available on the Internet at http://dnb.dnb.de.

ISBN 978-3-0356-2762-6

e-ISBN (PDF) 978-3-0356-2931-6

Im Westfeld 8, 4055 Basel, Switzerland

Part of Walter de Gruyter GmbH, Berlin/Boston

9 8 7 6 5 4 3 2 1

www.birkhauser.com